Technological Holocaust

(Edition Seven)

*A Call for the Heart of Humanity to Stand Up
& Save Itself From Technological Destruction.*

I0464707

by Sharon R. Poet

(Previously known as Namatari Neachi, Sharon Y. LaBree and Sharon Buck)

Other Books by Sharon R. Poet

The Heart Bud
Hints of Me in Poetry
Buds of Inspiration
Targeted in America (4 editions)
Embracing Sadness (Previously "Embracing Feelings)
The Personal Journals (book of 2005 booklets)
Technological Holocaust (7 editions)
Into the Light - Out of the Dark
Poetic Voice of a Targeted Individual (2 editions)
Ramblings of a Targeted Individual (12 editions)

Publications by Sharon R. Poet

(Previously known as Sharon LaBree, Sharon Buck and Namatari Neachi)

Calling All Hearts (2014)
Technological Holocaust (2012)
Public Notice (2012)
The Heart Bud (2009 to 2014)
Sharon's Bud (2007 & 2008)
The Personal Journals (2004 - 2006)

Poetic Publications
PO Box 383
Mont Vernon, NH 03057

Web Sites

www.PoeticPublications.com
www.heartbud.com
www.targetedinamerica.com
www.sharonpoet.com
www.sharonpoet-ti.blogspot.com
www.technologicalholocaust.com

Other Sites

www.heartbud.blogspot.com
www.targeted11.blogspot.com
www.technologicalholocaust.blogspot.com
www.sharonpoet-ti.blogspot.com
www.poeticpublications.com/zinfo.html

Contents

Its Worth Fighting For -

Humanity's Right to Freely Live and Love and Laugh and Cry.

Its Time to Fight Like Heaven!

Dedication

I dedicate this book to the Heart and Soul of humanity; to the Freedom of every nation and citizen around the globe. May the time soon come when our lives - our bodies, minds, hearts, spirits and natural process of personal growth, be protected from further technological and pharmaceutical harm.

Its Worth Fighting For -
Humanity's Right to Freely Live
and Love and Laugh and Cry.
Its Time to Fight Like Heaven!

Introduction

This is not a "theory." Its a fight for our lives. Its not a matter of if you "believe it" or not - its a matter of if you are aware and if you can care to help restore our safety and freedom.

I've struggled to expose criminal use of microwave weapons, since late 2011, with a very limited amount of resources available to me, and while being targeted and often forced to work on slyly infiltrated computers. Its been difficult to say the least.

I have had to rely primarily on my own experiences and microwave dulled instincts. And I ask you to do the same. Please read this book with an open mind and Heart - listen to your own instincts and Heart above all else.

There have been many previous editions of this book because my writings have grown with my realizations. . .originally rising from a desperate rush to save our crumbling lives. . . to what they are now - the best I can do to objectively share what I have experienced and witnessed. Some of the past editions may even have more information than this one. But they all merely grazes the surface of a global crisis, which is crying out for the Heart of humanity to realize and stop. I hope it inspires you to do some of your own research into the technological and pharmaceutical crimes that are being inflicted upon humanity.

I hope these writings inspire those who can do more to help bring an end to the suffering of Targeted Individuals and prevent the rest of humanity from falling into the trap of literal technological enslavement.

I ask that those who have viewed my earlier writings, please excuse me for possible areas of misinformation and periods of misplaced blame and anger. I, like many other Targeted Individuals, initially found myself rushing to expose this crisis under the oppression of threats, heavy technological targeting and being fed misinformation. It is indescribably difficult to do this while being targeted. . .and before full, honest, professional disclosure of these crimes. But I have done the best I can to combine and compare a scientific reports with my own insights and experiences. . .in order to come as close to it as I can at this point in time. I hope it helps all of us.

These writings began on www.poeticpublications.com and www.sharonpoet-ti.blogspot.com before shifting to www.targeted11.blogspot.com, www.technologicalholocaust.blogspot.com and then www.targetedinamerica.com and later also added to www.technologicalholocaust.com.

Give us STRENGTH, God to find our way
Through bullets hidden in microwaves.
And COURAGE, God to make a STAND
That saves our lives and FREEs our land.

If you look with your Heart you will see

5

Chapter One

Technological Holocaust

I am still deeply concerned that the covert "rescues" are actually enslavement and that technological modes of "protection" could be a sly enslavement performed by those who may also be placing filters in detection technologies, in order to prevent detection of the low frequencies that are used for mind control. Help spread the word on this.

Its Worth Fighting For -

Humanity's Right to Freely Live and Love and Laugh and Cry.

Its Time to Fight Like Heaven!

Tesla Technology
The Birth of a Technological Holocaust?

This is not a "theory." Its a fight for our lives. Its not a matter of if you "believe it" or not - its a matter of if you are aware and if you can care to help restore our safety and freedom.

Nicola Tesla was a visionary scientist - a man who was brilliant with mechanical/technological inventions, but also had the insight to look beyond the physical. He is reported to have had good intentions with his inventions and had even said, *"I prefer to be remembered as the inventor who abolished war. That will be my greatest pride."* But have there been attempts to hide certain parts of Tesla's inventions? And are those parts being used in a technological holocaust? I believe so.

The medical field, wealthy business men and governments, around the globe, were interested in his most popular inventions. His Xrays could be used to see into the human body...etc., his electrical inventions could be used to power mechanical equipment, his experiments with radio waves (microwaves) could be used for wireless communication, his "Death Ray" particle beam weapon could be used for defense in times of war...etc. This has all been historically recorded, along with his efforts to provide the world with free electrical power drawn from the energy in and around the Earth.

> *"The desire that guides me in all I do is the desire to harness the forces of nature to the service of mankind."* ~ Nikola Tesla *

On July 11, 1934 New York Times reported that Tesla invented the "Death Ray" - a particle beam weapon. But Tesla's experiments with trying to collect electricity from the atmosphere or Earth, appeared to have also proven that the weather could be manipulated and that living beings were seriously effected by the radio waves or "scaler waves," which he flooded the environment with. Other reports have stated that this process could drop a herd of cows or make birds and fish flee from the area...etc.

This appears to be where things began to turn into a dark direction. I feel that this dark turn was partly Tesla's fault, because he crossed a line that human beings should not ever cross; he aimed to play God - to technologically manipulate the energy in and around the Earth as well as in living organisms. **Technologies that can shoot radio waves into or around the Earth and living beings, in order to disrupt natural energy fields, is probably the most dangerous thing that humanity will ever be faced with.** Unfortunately these sorts of technological "accomplishments" fell into the dark closets of people who did not have good intentions. I think that, in the end, Tesla knew that his work would be used to harm humanity.

> "You may live to see man-made horrors beyond your comprehension." ~ Nikola Tesla *

Have you heard the news reports about a flock of birds suddenly falling out of the sky, or of a large school of whales beaching themselves, for no apparent reason? Have you noticed the unusual droughts, tidal waves and storms that have been happening around the globe? Have you wondered about the growing mental numbness and heartlessness in our world? Please start wondering and searching for True answers so that solid and perminant solutions can take place.

Tesla's more hidden experiments involved the effects of radio waves on the human brain and body. He revealed the fact that he did this when he talked about, *"...regular treatments of electricity, which I applied to myself. It puts into a tired body what it needs most - life force energy."*

But his was not the only body that he experimented on. Mark Twain made public statements about how much electricity Tesla had shot into his body without any ill effects.

Tesla's spiritual outlook, and the connections he made between physical health, spirituality and technological experimentation, probably baffled most people and scared some. He appeared to have begun deeply exploring this part of his work in England where he is reported to have joined up with Sr. William Crooks - a mystic who *"believed that human beings could communicate telepathically when they were attuned to high frequency brain waves."* Perhaps this is what Tesla was talking about when he said, "I would say a few words about that which fills my mind and concerns the welfare of all. I mean the transmission of intelligence..."

*"If you want to find the secrets of the universe, think in terms
of energy, frequency and vibration."* ~ Nikola Tesla *

After Tesla's visit to England he is said to have returned to the USA where, he retreated from society, and continued private experimentations. These experiments are probably the ones that more deeply involved the effects of remotely projected radio waves (microwaves) on the human brain and body. Ironically, it was during this time that his lab was destroyed in a fire. Did the fire destroy his work or was the fire set to hide the theft of it? I feel that it was set to hide the theft of it.

Tesla had connections with influential people around the globe. Supporters for his experiments included people like J.P. Morgan and were probably not a secret to the type of people who were capable of utilizing them in ways that could inflict harm. Can you imagine the effects of this sort of technology being in the hands of the type of people who want to inflict discord and suffering or in the hands of people who want to perform remote technological brainwashings or mind control on fellow human beings?

*"Our entire biological system, the brain and the earth itself work on the same frequencies.
If we can control that resonant system electronically we could directly control the entire
mental system of human kind."* ~ author/phylosopher, Velimir Abramovic

Please look around the world - look with your Heart and listen to your instincts and you will see the effects of mind control technologies. They are around us and within us. They show up in the growing lack of Heart that is surrounded by increasing levels of selfishness, greed, immorality and crime. We are experiencing the results of a lethal, secret Technological Holocaust, which most people are not aware of due to their lack of knowledge of the technologies and their effects on the human body and brain. This crisis is not "on the horizon." It has been happening since at least 1934, possibly as early as the 1890s. And it also includes inconspicuous eugenics based targetings.

*"You can induce virtually any effect that a chemical can cause in a living system, with an
external primarily extremely low frequency electromagnetic field"* ~ bioengineer Eldon Byrd

**Criminal use of radio wave technologies has created
a holocaustal situation that must be stopped ASAP.**

I feel that those who took Tesla's mind control test results began infiltrating government agencies and other organizations where they could be in positions of power. . .and then used the technologies to interfere with the brains of their colleges. . .rendering them too numb to not go along with them. I even believe that, as people began to realize the targeting, things were often manipulated in ways that placed blame on other things (like aliens) or their primary targets. (These targets have been countries, companies or people whom they want to take over or destroy.) I believe that they had even set up a "protective" technological measure, which actually slyly finishes enslaving people and then providing filtered detection technologies, in order to keep the technological mind control hidden.

I believe that these technologies were being used in Germany during the second world war, which explains the levels of control Hitler had over people. It even appears that Hitler himself may have been being targeted with pharmaceuticals as well as radio waves.

There is an interesting "Death Ray" connection between Nikola Tesla's work and the research of Bernhard Schreiber in "THE MEN BEHIND HITLER - A German warning to the world." In this book Bernhard Schreiber exposed the field of psychiatry as being connected to an ongoing holocaust. He wrote, *"Hitler was an evil man and no one would want to assert that he was not responsible*

8

for the things that happened in Germany, but in blaming Hitler for all the evils one is over-looking a considerable number of those who are truly responsible, people who are being allowed to pursue their course to similar ends all over again - nothing to stop them."

In another part of his book, which appears to have been erased, Schreiber had spoken of Dr. Paul Joseph Goebbels bragging, in 1944 - after Hitler's Reich had fallen, about a secret weapon - a "death ray" that would lead to the "rebirth of the Reich." The fact that this statement was erased has great significance, because it shows us what is being hidden. (Sometimes we can learn more by watching what is being erased.)

Schreiber's conclusion, where he tried to figure out what the weapon was, remains at the end of his book. He wrote, *"The Nazis may have been disbanded, but the psychiatrists still linger on among us. Maybe this is the secret weapon Goebbels boasted about which would lead to the rebirth of the Reich - not a super-bomb and not a death ray, but a blueprint for a psychiatric slave state."* He was right, but obviously just was not familiar with Tesla's inventions. Those "Death Rays," are physical weapons that, among other things, are called particle beam weapons, microwave weapons, psychotronic weapons. . .and are being used to target humanity.

Men Behind Hitler - A German Warning to the World
www.targetedinamerica.com/menhitler.html

The technological mind control part of the targeting is most effective on people who are taking certain types of chemicals and drugs, including psychiatric drugs, which have been being forced into masses of people through food and water. They have also been heavily pushed onto people through the medical field. www.targetedinamerica.com/mindcont.html

I believe that there is a direct connection between technological mind control targeting, false mental illness diagnosis and the administering of psychiatric drugs. Humanity truly has been being reduced to that, "psychiatric slave state," which prevents personal and spiritual growth and aids the process of technological mind control.

Dr. Thomas Stephen Szasz was a pioneer in the movement to help us and save the field of psychiatry from the shame of unethical "mental illness" labels and the harmful medications that they justify. He said, *"The task we set ourselves to combat psychiatric coercion is important... Its a noble task - a task in the pursuit of which we must, regardless of absticles, persevere. Our conscience commands that we do no less."* Dr. Szasz wrote a book called "On the Myth of Mental Illness" and says that "there is no such thing as mental illness." Szasz says that "Labeling a child as 'mentally ill' is a stigmatization, not a diagnosis. Giving the child a psychiatric drug is poisoning, not treatment." I was so glad to have found Dr. Szasz. Thank God for his courage and insight.

The Psychiatric Connection
www.targetedinamerica.com/psychiatry.html

That "Rebirth of the Reich" appears to have begun a long time ago. In this holocaust remotely inflicted technological mind control is being performed with a lethal silence, which prevents personal and spiritual growth and pits citizens against governments and governments against citizens as well as family members against each other...etc.

I feel that the technologies, which can direct radio waves over long distances, have advanced into satellite type stations, which is how we can get internet access from satellites. (I had a dream, and an experience, which suggested that satellites are being used to target us.)

Aside from mind control and eugenics based targetings, technological weather modification and earthquake inducing lasers have been being used as weapons against humanity. This is all in desperate need of being exposed so that countries and people can understand and pull together to support each other until our Freedom is restored.

**Never, in the history of humanity, has there been a more
critical time for the Heart of citizens and governments
to pull together in a stand for Freedom.**

Though the end, of the first video I'd placed here, only blamed the ground based HAARP technology, there are at least 5 similar technologies placed around the world. At least one was built long before HAARP. There are many ways that radio waves can be directed at people, communities and countries. Any technology, that can direct radio waves, can be criminally used. And criminal use of them must be stopped as quickly as possible. (www.youtube.com/watch?v=ioKb_6nP3bQ) This video was erased shortly after I put it up, but the one below popped up.

Nikola Tesla: The Full Life Story 2015 (History Channel Documentary)
https://www.youtube.com/watch?v=fXTLn9rYcf0

Nikola Tesla | The Missing Secrets | Full Documentary
https://www.youtube.com/watch?v=1b5ciAZmyd0

Nicola Tesla could have done so much more to help humanity, had his inventions not ended up in the wrong hands. . .and had he not aimed to harness/disrupt the vital life force energy, which we need to remain natural and undisturbed, in order for our world, including humanity, to thrive and grow and heal and evolve into all that it was meant to be.

Aside from the lethal targeting there is still cause for concern. Some scientists feel that it is the types of technologies, which shoot radio waves into the ionosphere, that are causing global warming. Even too much positive use of them can damage the Earth and its atmosphere. And many scientists have proven the harm that various frequencies of radio waves can have on human beings and animals. There is cause for concern about our environments being flooded with radio waves for wireless internet, which some people are sensitive to. There are valid concerns about the effect of even simple things like cell phones and computers on young children. But far worse than this is happening. And communities being flooded with wifi may even prevent accurate detection of the frequencies that are used for mind control.

Please do your own research, but more important. . .listen to your Heart and instincts above all else. Concrete proof may be hard to find. If you feel things to be true, they probably are. If you blindly disbelieve, you have probably been brainwashed. Either way, please do all that you can to help regain our freedom. And please keep in mind that those who target us have had a pattern of manipulating people into placing blame where it is not due. Please help bring a peaceful end to this holocaust.

I beg the good in our governments to quickly expose this holocaust and prevent further criminal use of the pharmaceuticals and technologies that have been targeting them as well as us. Please give people the opportunity to understand and pull together to support each other until the crimes can be stopped - please give Love the chance to conquer the holocaustal stealth of hate. Please help break the silence. It is hurt us

If criminal use of the technologies can not quickly happen, I pray for the technologies to be destroyed, in order to save the Heart and Soul of humanity as well as the lives of the people who are now being either physically or psychologically destroyed. God help us all.

<u>On a Spiritual Level</u>

I have had strong feelings about Tesla experimenting with shooting radio waves into his own body, in order to induce spiritual states of mind. I have no doubt that this boosted his energy and perhaps even lead him into euphoric experiences. But my gut feeling is that this process is merely a temporary 'quick fix' type of thing, which can even prevent the vital process of natural spiritual growth.

<u>I strongly feel that deeper healing (spiritual/personal growth) cannot be done technologically and that doing this can throw a person's natural energy field off balance and actually inflict harm. Radio waves shot into the human brain can seriously interfere with our natural ways of thinking, feeling and healing. It hurts us. Sometimes it even damages us. Technologically induced euphoria, energy or spiritual healing bypasses the most important part of the natural healing process, which is needed for the growth of our Souls - it bypasses our Hearts.</u>

I feel that Tesla suffered from the negative side of his inventions through the results of his experiments on himself and possibly also through those who had stolen his inventions and then utilized them on him - shooting disruptive radio waves at him, literally driving him crazy in the final decades of his life. I look at the pictures of Tesla and feel sad, because they depict a brilliant, eager young man who ended up shrouded in darkness.

During the early years of my own spiritual searching and focusing on various methods of spiritual healing, (in the 1980s) I gained insights on how critically important it is for us to focus primarily on healing our Hearts. . .and that in healing our Hearts everything else automatically clicks into place. I believe that this is what Jesus had aimed to teach, which is why He is often depicted with Light shining from His Heart. That Love/Light is the most vital thing in life. It is the center of life itself and Heart is its conductor. This is even depicted in the eastern chakra system where Heart is at the center. If the healing of the Heart is bypassed, in the process of spiritual growth, it creates an imbalance. Actually all we really need to do is heal our Hearts. Heal the Heart and all else automatically begins to heal in a good and balanced way. This is a True and lasting method of spiritual healing and personal growth.

When we are free from the radio wave targeting, which blocks our Hearts, and can focus on healing and opening our Hearts, we will find that the potential of human beings soars beyond what most people can even imagine possible. With Heart/Love freely leading our lives we can create literal Heaven on Earth. Lets do it.

On a More Personal Level

I could relate to some of Tesla's experiences, like the loss in a fire, being plagiarized and made destitute by greed, being shunned and ridiculed, turning to the birds for the love that is too severely lacking in my life, and futuristic dreams and visions that lead to obsessions with helping humanity and making our world a better place.

When I was a child I felt like there was something horribly wrong in the world. And I had a vision of the perfect world - one that was filled with Love and compassion and no wars...etc. I had a vision of a better world, which has remained in my heart and kept me feeling dissatisfied with the way things are. I wanted it to be better. People used to tell me that I was "too idealistic." But how can things ever get better if we do not have a more ideal vision to aim for?

Later in life, starting in the mid 1980s, I had strived to face life's difficulties, and release my suppressed pain, through feeling and healing my own heart. I later aimed to publicly sharing my experiences and healing process, in order to help others to heal their Hearts. I had felt a deep calling to do this.

I feel that the more open our Hearts are, the closer we automatically get to God/Love/Light. And this is what my work is all about - bringing more Love into myself and into our troubled world. In 1999 I had completed a book, ("Embracing Feelings") with this focus. But the final manuscript was destroyed in a suspicious fire. I have not been able to fully reconstruct this book. I have a new version of it, but it's not the same. Since the end of 2005 I have been being too heavily targeted to even continue my work effectively. My work has been being sabotaged and my life has been being slowly destroyed by ongoing technological targeting, chemical targeting and psychological harassment. There have even been efforts to slander me, frame me...etc. I am trapped in an indescribable hell.

I have wondered if other people, who have a similar life purpose, are also being heavily targeted. Are most of the long term heavily "Targeted Individuals" the ones who have had a vision of a better world and a way that they have wanted to help create it?

Since 2006 I have been forced to focus on surviving, exposing and stopping the targeting, so that I can freely think, feel, live and work. I need that so badly that every day of remaining in this torturous prison feels too long. I pray for Freedom and hope it soon comes for all of humanity. . .me too.

P.S. Please excuse my bloops and blunders in my targeted reconstruction of this article. I have been hit hard since the end of 2014 when I started pulling this material together. My earlier post on the Death Ray; Dr. Geobbels's Mystery Weapon; http://sharonpoet-ti.blogspot.com/2014/12/dr-paul-joseph-geobbelss-mystery-weapon.html

The first full article, that I put on the web, was altered. Now, in March of 2016, as I aim to reconstruct it, I am still experiencing interference. For three days I worked on it while being hit with painful laser weapon shots, disruptive radio waves being blasted into my brain and noise campaigns. On two occasions my computer was remotely shut down and parts of my work wiped out. Hopefully what is left of it will help us. I could write a whole book on this, but I can not do it while being held captive in this invisible prison that watches me, judges me and technologically attacks me. I wish I could. I wish I were free. I wish we all were.

* These quotes came from https://teslauniverse.com/nikola-tesla/quotes. Even IF they are not completely accurate, they surely do depict Tesla's true focus and intentions.

Give us STRENGTH, God...to find our way through bullets hidden in microwaves, and COURAGE, God...to make a STAND that saves our lives and FREEs our land.

Connecting Dots Between False "Mental Illness" Diagnosis, Psychiatric Pharmaceuticals, Microwave Targeting and Eugenics

If you, or anyone you know, are taking antidepressants or other types of psychiatric drugs you should read this article and listen to the depths of your own wisdom above all else. The doctors do not always "know best." Please educate yourself and gain the opportunity to make the healthiest choices for yourself and your loved ones.

Is it a coincidence that the field of psychiatry is suspected, in Bernard Schreiber's research, to be involved in a continuation of holocaustal crimes against humanity since Hitler's holocaust. . .that, according to General Barrie Trower, microwave weapons began perfecting their ability to remotely inflict symptoms that mimic "mental Illness" in targeted individuals in the 1950s. . .that the field of psychiatry was making shifts from Freudian methods to medicatable "mental Illness" diagnosis in the 1950s. . .and then (also in the 1950's) along came people like Dr Thomas Stephen Szasz to stand up for humanity against false "mental illness" labels?

In my first edition of "Embracing Feelings" I strongly encouraged people to work out personal issues through embracing/releasing suppressed feelings, instead of taking harmful medications. The first edition of this book was completed in 1999 and was followed by severe rounds of targeting which ended in the loss of 3 pets, loved ones and a suspicious fire that raged through my New Hampshire home...etc.

I am not sure if this round of targeting happened solely because of my book. It does appear to be due to other things also. But, the final manuscript, along with most of my other writings, were lost in the fire. And due to how much interference I had while reconstructing it, it was obviously part of the reason for vamp ups in the targeting. But who would want my work destroyed? And why?

My work, was not offensive or threatening - it was all about healing the heart of humanity and bringing more Love into our troubled world. The only types of people, whom I've thought could be apposed to it, had to be some sort of dark occult. But could it have been a threat to some other sort of organization?

My drive to encourage healing instead of medicating, grew from my own experiences with personal growth and things I'd witnessed in a mental health facility, which I'd worked at. I found that through embracing our feelings we can heal from past trauma and that medicating our issues prevents resolution and stalls or prevents personal growth.

I have never taken psychiatric medications but I know people who have and I've witnessed the stagnation, and sometimes even harm, that they can inflict, especially on people who are not really "mentally ill."

Even without meds, the "mental illness" labeling can destroy a person's life. I believe that there are many children and adults who were being labeled although they are perfectly healthy and just needed to be encouraged to allow a healthy grieving process or an outlet for their frustration...etc. I'd not mentioned this in the book, but its time to.

People who are labeled as mentally ill can lose their rights to take care of themselves and make decisions for themselves. Their work and/or homes can be taken over by relatives. In the worst of situations they can be forced into institutions and medicated against their will. This form of destruction of healthy human beings is probably the worst kind of crime that humanity has ever been faced with, because most victims are not aware and the full scope of the crimes are not yet openly exposed and recognized by the general public.

I'm not a doctor. But, when I wrote the book, I had a lot of experience with the embracing of feelings, in order to heal from past trauma, instead of suppressing it with medications. I personally trust the wisdom of experience and instincts above scientific text book knowledge, when it comes to personal growth, because every individual's circumstances are different. But, for those who need a Dr. label, in order to believe that we have a serious problem here, I've compiled statements from doctors and other scientific experts.

I feel that the crisis humanity is faced with has multiple dimensions which are summed up in the first paragraph of this article. Some are so obvious that investigations are probably already taking place. And some may be difficult for you to believe, especially if you are not aware of the capabilities of microwave and psychotronic weapons. Please become aware.

The following doctors have waged a war against the psychiatric label of "mental illness" and the harmful pharmaceuticals that are either prescribed for or forced upon victims of the label, which comes from the DSM - the book which defines criteria for the diagnosis of "mental illness." <u>Some psychiatrists are even concerned that the labels and psychiatric drugs are so unethical that it could destroy the credibility of the whole psychiatric profession</u>.

In the video below Dr. Paula Caplan explains that, *"the whole process of mental illness diagnosis is unregulated. . . there are categories in the DSM that have scientifically proven not to exist and they're in there anyway."* Normal *"things that everybody goes through. . . get diagnosed as "mental illness" for example; if someone close to you dies and your still depressed two weeks later, now, according to the*

13

DSM you have minor depressive disorder" And she expresses a VERY justifiable concern that normal feelings and social behaviors, especially those of natural grieving or anger, are listed in the DSM as cause to label a person with "mental illness." I like her approach and was glad to find her video, although it has changed addresses since. I feel that the false labels and harmful medications soar beyond unethical and into what can be called the destruction of our natural process of personal and spiritual growth. (Dr. Paula Caplan wrote a book entitled, "They Say Your Crazy.")

Dr. Paula Caplan on the dangers of false "mental illness" labels
https://www.youtube.com/watch?v=dwxEdRYQqCU

This is not my video, but it is my recording of it, due to its address being changed. Hopefully it will be allowed to remain at this address. Its contains VIP information by a woman whom I believe may be being targeted.
Previously posted videos;
https://www.youtube.com/watch?v=sZpr8sVih70
https://www.youtube.com/watch?v=nKbybLM12yc)

I was shocked to find these headlines, which read, *"I think my child is Mentally ill"* on top of a picture of a sad little girl who has the words *"I feel sad,"* printed above her head. This was on page 10 of the the November, 2015 issue of the New Hampshire "Parenting" magazine. And, to me, it is a blatant reminder of the unhealthy push for false mental illness diagnosis and it's treatment, which often includes psychiatric drugs and other methods of suppressing feelings, instead of healing.

These sorts of ads come across as brainwashings aimed to make people think that sadness is wrong or worse - that it should raise concerns about "mental illness." These sorts of ads should raise red flags in those who care about children and the future health and safety of humanity.

Sadness is one of the natural emotions we were born with and we were born with the capability of feeling it for a good reason. When encouraged, sadness is part of a healthy grieving process, which helps us to release our pain. When we do not allow this process of feeling and releasing our pain our hearts become blocked and this is not good for us. Actually, its harmful for natural feelings of sadness to be suppressed and blocked, instead of felt and healed, especially in children. And this is ASIDE from the dangers of these drugs aiding technological mind control.

Dr. Thomas Stephen Szasz appears to be the pioneer in the movement to save the field of psychiatry from the shame of unethical "mental illness" labels and harmful medications that they justify. On the battle to save humanity from harmful pharmaceuticals Szasz says, *"The task we set ourselves to combat psychiatric coercion is important. I think its. . . a noble task - a task in the pursuit of which we must, regardless of obstacles, persevere. Our conscience commands that we do no less."*

Dr. Szasz wrote a book called "On the Myth of Mental Illness" and says that <u>"there is no such thing as mental illness."</u> Szasz says that *"Labeling a child as 'mentally ill' is a stigmatization, not a diagnosis. Giving the child a psychiatric drug is poisoning, not treatment."*

Site raised for Dr. Thomas Stephen Szasz
www.szasz.com

Dr. Colin Ross has extensively studied scientific data which disproves the validity of "mental illness" labels and renders most of the prescribed drugs harmful or ineffective in most patients. He and other ethical psychiatrists strongly feel that the DSM's false labels, need to be exposed and stopped. He says, *"Its not going to stop until psychiatry takes responsibility for what its doing and the public gets educated and says wait a minute I'm not buying this anymore."* (Dr. Colin Ross wrote a book entitled, "The

14

Great Psychiatry Scam.")

Is Psychiatry A Scam? Truth About Mental Disorders, Psychiatrists Colin Ross and Corrina on Psychetruth
https://www.youtube.com/watch?v=AG1VHpsgUcY

How a Mental Health Disorder is Discovered: DSM Truth, Psychiatrist Colin Ross and Corrina on Psychetruth
https://www.youtube.com/watch?v=2jFdBIZLwqU

Dr. Peter Breggin makes a strong stand against the psychiatric labeling and pharmaceuticals on an interview with FOX News in the videos below. Breggin calls the psychiatric drugging *"medication spell binding"* and that they do *"more harm than good."* His focus is more on the violence that drugs can instigate in victims.

There is, however, a danger that those, who are not aware of the technological part of the targeting, can easily blame psychiatric pharmaceuticals for severe cases of technological targeting like in the Navy yard shooting, which Breggin blames on a pharmaceuticals. I feel that it was both. The shooter was obviously being targeted.

FOX News ; Peter Breggin MD Psychiatric Drugs Part I to III
https://www.youtube.com/watch?v=3IXUOnn5PiQ
https://www.youtube.com/watch?v=WJNMBr2zRX8
https://www.youtube.com/watch?v=HaPFmgLNwYs

Dr Jeffery Schaler went so far as to say, *"The diagnosis of mental illness is always a weapon."*

https://www.youtube.com/watch?v=BhC6hUZJIJ0

Surprising Research into Psychiatry and Eugenics

Bernard Schreiber, wrote a book entitled "Men Behind Hitler - A German Warning to the World." His research exposes the profession of psychiatry being involved in a covert continuation of the eugenics movement. He wrote, *"Hitler was an evil man and no one would want to assert that he was not responsible for the things that happened in Germany, but in blaming Hitler for all the evils one is overlooking a considerable number of those who are truly responsible, people who are being allowed to pursue their course to similar ends all over again - nothing to stop them."*

In another part of his book, which appears to have been erased, Schreiber had spoken of Dr. Paul Joseph Goebbels bragging about a secret weapon - a "death ray," which would lead to the "rebirth of the Reich." I believe that the weapons Goebbels spoke of are microwave weapons.

Schreiber's book is *"Humbly dedicated to the memory of countless ordinary people who's lives were taken because they were considered less than perfect, and therefore, unworthy to live."*

Free Download of "Men Behind Hitler..."
http://www.goodreads.com/book/show/6011376-the-men-behind-hitler
(Also find it here; www.targetedinamerica.com/menhitler.html)

The History of Psychiatry

The following video shares part of the history of psychiatry. Potent quotes from this video are, *"Far from being places of healing, psychiatric institutions have always functioned as the worst sort of prisons. . .where someone can be incarcerated against their will without even being charged with a crime."* and *"This is an industry that doesn't just specialize in destroying people's lives, it kills their souls."*

Age of Fear: Psychiatry's Reign of Terror
https://www.youtube.com/watch?v=YA_MwaRLzm8

And Here's the Most Dangerous Part of the Puzzle

Microwave expert, General Barrie Trower describes microwave weapons , and their use on human beings since the 1950s. On the subject of the use of microwave weapons on human beings Trower says, *"By changing the pulse frequency... of the microwaves going into the brain and interfering with the brain. . . you could induce psychiatric illnesses to the point where a psychiatrist could not tell if it is a genuine psychiatric illness or an induced psychiatric illness."*

He describes these weapons as technologies which shoot beams of microwaves into space to bounce off of the ionosphere and be redirected to the target, which can be virtually anywhere in the world. His description fits technologies like the Russian SURA and the American HAARP. (At least six of these types of technologies are reported to exist around the globe.) This sort of targeting would also have to be done in conjunction with satellite surveillance systems. So, there are many technologies used.

Microwave Expert - General Barrie Trower
https://www.youtube.com/watch?v=kvn-8lTy0oc

According to experts the mind control parts of Microwave Weapons are most successful on people who are taking mood altering drugs, like anti-depressants.

Aside from the book I wrote in 1999, there is one more thing I did before hell broke loose around me. I'd had a prophetic dream, which showed criminal contamination of a public water supply and had shared my concern about it with a few people, including the water department. But many difficult years passed, as I struggled to survive the targeting, before I found the following information and realized that this is probably what my dream was warning of.

In 2008, news reports stated that pharmaceutical drugs, including antidepressants, were being found in around 24 major public water supplies throughout the USA.

16

Drugs found in 24 USA public water supplies
https://www.youtube.com/watch?v=LYD-qf9pZAg

In 2013 a secretary at a New Hampshire Environmental Protection Agency told me that these drugs are STILL being found in our public water supplies! Reports say that its from the "run off" of pharmaceuticals that are improperly disposed of. Perhaps those who are coming to this conclusion are not aware of the microwave and pharmaceutical targeting of humanity?

Because of the dream I had about criminals putting toxins in public water, and the fact that I had this dream while I was focused on writing a book to help steer humanity toward healing, instead of medicating. . .I believe that the drugs are being placed into the water by groups of criminals who are involved in a covert continuation of the eugenics movement, which includes efforts to control human beings with microwave technologies, which are aided by mood altering drugs. Whether the drugs are from "run off" or intentional contamination the resulting problem remains the same and something needs to be done to stop it.

I understand how unreal this all sounds. But PLEASE think about this and DO NOT STOP thinking about it until you feel/know the absolute Truth. Too much is at stake for it to be washed away with blind disbelief.

Even when we put aside the unpublicized technological part of these crimes, the process of falsely labeling, and inflicting healthy human beings with harmful pharmaceuticals, are still holocaustal crimes against humanity, especially when those human beings are rendered too numb to freely live their lives or when they lose their credibility and right to make their own choices with their own lives.

But please do not push aside the technological part, because it is the worst of it - it is what makes the rest succeed - it is what makes it all so difficult to face - it is the part that is inflicting humanity with brainwashings and more severe levels of mind control as well as other lethal types of targeting. Criminal use of microwave weapons is the a critical problem that must be faced and stopped as quickly as possible.

Why would people be targeted in ways that could make them go see a doctor and get labeled as "mentally ill," and either prescribed or forced to take, harmful pharmaceuticals? Who would put drugs in our water? Who benefits from this destruction of lives? I believe that the roots causes are greed and evil. And we ALL need to do what we can to stop these crimes and save humanity from further destruction. Find more on www.targetedinamerica.com

Lets face this with determination to
Preserve humanity, with gentle hands
Reaching out to those in need, with
Peaceful non-acceptance of the evil seed.

ITS OK TO FEEL SAD, ANGRY OR SCARED. THESE ARE NATURAL HUMAN FEELINGS THAT WE NEED TO EMBRACE AND WORK THROUGH, IN ORDER TO GROW INTO ALL THAT WE ARE MEANT TO BE.

17

P.S. What are the dangers of so many people being pushed to take psychiatric pharmaceuticals for having natural human feelings and issues? I feel that there are many. . .some of which have probably not yet been realized by even the field of science. Tampering with the brain of human beings interferes with the natural process of personal and spiritual growth, which I feel is what life on Earth is all about. The damage that is being inflicted upon people is immeasurable. And some situations are even worse - some people are being literally tortured. Many have reported being targeted with microwave weapons, laser weapons, mind control technologies....etc. These people are commonly known as "Targeted Individuals." I have heard that many of these victims have been forcefully labeled as "mental ill" instead of being protected from further abuse. The destructive and discrediting "mental illness" labels are just part of the horrible crimes that are being committed against heavily targeted people.

I am a victim of technological targeting and have had to fight against the "mental illness" label which I have not been inflicted with, but certainly not out of their lack of trying to shove me in that direction. I am fighting to not become another labeled pharmaceutical victim on top of all the other parts of the targeting, which are more than any human being should have to live with for ANY length of time. It is Truly insane that any human being would have to fight to not be falsely labeled as "mentally ill" or force medicated by criminals who are also targeting us with microwave weapons! The hell that we are going through soars beyond inhumane. The targeting has sabotaged nearly every aspect of my life. I used to own a nice country home and now I live in a vehicle as I fight to expose these crimes. Many others are experiencing similar levels of targeting. We are suffering indescribably.

Its taken me a while to pull this article together! I was shot with painful doses of microwaves when I first found Dr. Szasz on the web in December 2014. I was not able to start fleshing out this article until around February 5, 2015 and then there appeared to be an aim to prevent me from posting and printing it as I continued to be heavily targeted with microwaves.

Finding validation for the psychiatric crimes was a huge relief for me and it gives credibility to information I've been sharing, and have often been degraded for, for decades. I hope it helps you to not only realize what is happening, but to also make the healthiest of choices for yourself and your loved ones. Its OK to have natural human feelings. they are actually a prerequisite for personal growth. And we all need to be growing and evolving into all that we are meant to be.

"The task we set ourselves to combat psychiatric coercion is important. I think its. . . a noble task - a task in the pursuit of which we must, regardless of obstacles, persevere. Our conscience commands that we do no less." ~ Dr. Thomas Stephen Szasz

Technological Holocaust

Criminal use of satellite surveillance systems, and other radio wave transmitters, is a serious crisis, which may be the most dangerous thing humanity will ever have to face.

And we MUST face it in order to stop it.

Most people are aware that radio waves can be sent down to their homes or computers, from satellites, for the purpose of internet access. But most people are not aware that radio waves (also called Microwaves) can be criminally used.

Condensed beams of radio waves can be shot into the human brain and other organs, in order to effect their functionality in ways that range from mental numbness, brainwashings and physical illness . . .to brain damage and death. And unfortunately, this is happening to people.

Never, in the history of humanity, has there been a more crucial time for our HEARTS to rise into a peaceful fight for humanity's safety and freedom.

This site aims to help bring public awareness to a crisis to a Technological Holocaust, which lurks in the shadows of a Covert War. Though there is some controversy about if this is an actual holocaust or not, I have decided to retain the title, because I feel that it is.

It is not possible to accurately determine how many people have been harmed, at this point. In my opinion, just one person psychologically mutilated or covertly murdered with what has been classified as "non- lethal weapons" is too many.

But in the tiny corner of it that I have directly experienced and witnessed, there are dozens of victims, and most of them remain unaware, due to a lack of awareness of the technologies that are being used on us.

Criminal use of radio wave technologies is a serious threat to ALL of humanity. As you face this devastating reality, your mind may not want to believe it and your Heart may be struck with a pain that its not going to want to feel, but I hope you find the courage to face and feel it anyway, because out of those feelings, may be berthed a new remedy that has no other way of existing. We need that remedy. The dangers in ignoring criminal use of radio wave technologies, and allowing them to continue and grow, can not be under stated.

What are the technologies and how are they used?

I'm not a technological expert. My experience/wisdom is in being a long term victim of criminal use of Microwave and Laser Weapons, as well as a witness to various targetings of dozens of other people. But, due to my experiences, I can give you an introduction to this crisis, with the hope that you will do more research and strive to help expose, and bring an end to, these horrible crimes against humanity.

That I know of there are five different types of technological targeting. 1. Remotely inflicted behavior modification or brainwashing on NON-consenting human beings. 2. Remote technological torturing of Targeted Individuals in ways that intentionally inflict physical and psychological distress. 3. Remote technological medical experimentation on NON-consenting human beings. 4. Eugenics based microwaving that inflicts terminal ill-nesses like lupus, leukemia, heart attacks, lung diseases, diabetes, tumorous cancers, adrenal gland failure, kidney failure...etc. 5. Criminal use of weather modification tech-nologies and laser weapons, which can instigate lethal storms as well as earth quakes and possibly even volcanic eruptions.

Technologies like the Russian SURA, the American HAARP, Gwen Towers, Cell Towers and Satellites are the primary suspects in these technological crimes. Most peo-ple seem to be blaming cell towers and those things that look like cell towers, but are reported to be more than that. (please excuse my ignorance. I'm being microwaved. ;-)
I feel that there are probably other technologies, which most of us are not aware of, certainly many that I am not aware of. (But, the obvious Truth is that ANY technology, which transmits radio waves, can be criminally used.)
Due to what I've witnessed and experienced, I feel certain that at least SOME of the technological targeting is being done VERY remotely. I have been being targeted with various types of microwave and laser weapons, which have the capability of quickly cir-cling around and coming at me from the sky, and from different directions and angles, when I am in areas where there are no other houses, vehicles, cell towers...etc. The only technologies, that I know of, which can do these types of maneuvers, are satellites and technologies that bounce radio waves off of the ionosphere to be redirected.

Among the names and applications of these technologies are; Information Weapons, Psychological Warfare, Psychotronic Weapons, Microwave Weapons, Behavior Modification Technologies, Synthetic Telepathy, Acoustic psycho-correction, Satellite Terrorism, Remote Neural Monitoring, Voice of god Weapon, Radio Frequency Weapons, Bio-communications technologies, Radio Wave Mind Control, Electroenergetics, Electronic Harassment, Directed Energy Weapons, Brain Warfare, Geophysical weapons, Mind Machine, Psychological Language Machine, Acoustic Heterodyne Weapon, Embryonic Holography, Optogenetics, Electromagnetic Stalking, Bio-com-munications Technologies, Radio Wave Mind Control, Microwave Mind Control, Bio-Electromagnetic Technologies, Electronic Harassment, Directed Energy Weapons, Brain Warfare, Bio-energetics, Electromagnetic (EMR) mind control, Electroenergetics, Geophysical Weapons, Psychoneurological Weapons, M.I.N.D. - Magnetic Integrated Neuron Duplicator, Psychic Warfare, Acoustic Heterodyne...etc.

There is scientific proof of the existence of these technologies and the fact that their criminal use has been kept secret is of concern to many.
Experimentation with the effects of radio waves (microwaves) shot into various parts of the brain and other organs are reported to have begun by the 1970s. Some reports state that the British were perfecting the technologies in the 1950s. World War Two victims of Nazi concentration camps are reported to have been experimented on with microwaves. In 1965 the American New York Times published an article on the remote mind control experimentations of the famous Dr. Jose Delgado. Part of the MKULTRA mind control program was exposed in the 1970s. Japan and other technologically advanced countries and organizations are sure to have done some of their own.
MOST of the microwave experimentation on human beings has probably never been publicly shared, because it is too criminal and inhumane for public approval.
Most of the published information appears to be about the USA and this leads people to blame only the USA for the targeting. But in my mind this merely proves that the USA

has been more open about it. . .and therefore probably less guilty. I am not saying that America is perfect, but I believe that America has been being slandered and infiltrated and struggling to retain its Freedom.

My experiences and insights tell me that there are at least two levels of targeting happening. One seems to be a long term covert effort to take over the USA and the Liberty its citizens have fought to retain.

I believe that the USA is being infiltrated and that some of our own civilians, military and government personnel are victims of remotely inflicted mind control, and that brainwashed victims have been being recruited into a deceptive program that is now aiming for complete control of the USA and its citizens.

Through this process, it appears that USA citizens and government are being pitted against each other. The manipulations to inflict discord and distractions are many.

I pray that, through the exposure of the technological part of these crimes, the Hearts of USA citizens and government officials will unite into a PEACEFUL stand for Freedom, and then help other countries do the same, because it appears that the long term goal of the technological targeting seems to be complete control over humanity.

Another level of the targeting appears to be being performed for experimentation and also as literal satanic types of torture. I'm not sure if it is all being done by the same group, but it appears to be.

Scientists and researchers have reported that, since at least the 1960s, behavior modification programs have been being implemented, around the globe, through remotely directing beams of encoded radio waves into the brains of individuals or into whole communities and countries, in efforts to accomplish brainwashings. Though most people seem totally unaware of this, it is not a new occurrence.

It appears that USA President John F. Kennedy tried to expose this crisis in 1961 in a speech he deliver to the press. There appeared to be an attempt to expose and stop criminal use of radio wave technologies, in the 1970s, when a small part of MKULTRA and Russia's microwaving of the US embassy were exposed. Apparently, it was thought to have been stopped. But I, along with many others, feel certain of the sly continuation and growth of these technological crimes.

JFK Secret Societies Speech (full version)
https://www.youtube.com/watch?v=zdMbmdFOvTs

I feel that the most dangerous part of the technological targeting is remotely inflicted mind control, because it robs us of the psychological freedom which is needed for our natural process of personal and spiritual growth.

Most mind control victims seem to remain completely unaware of what is happening to them. They think all is fine, but are not being allowed to freely follow their own thoughts, feelings and instincts. Some victims are aware and are fighting for their lives while being physically and/or psychologically tortured. Some are terrified and do not know what to do. And some have been psychologically mutilated to the point of ending up with permanent brain damage. When I think of them, it burns, like a painful fire in my heart. Its just too horribly wrong! And I pray that these lethal crimes can be realized and stopped before too many more people are hurt, and before humanity, as a whole, completely loses the personal Freedom that is need, in order to grow and evolve into all that we are meant to be.

Please think about this long and hard. This is Truly the most dangerous thing humanity has ever had to face. . .and it MUST be faced, because it has already been happening for too long and is already harming too many. Perhaps this next statement can help you to realize how wide spread this crisis is. . .

21

Around the year 2000 my concern shared a prophetic dream, which showed criminal contamination of a public water supply. By May of 2001, most of my pets were dead or missing, my daughters and I were suddenly surrounded by at least 5 unusual deaths, unbelievable levels of chaos and a suspicious fire in my home, which destroyed most of my writings...etc. Pieces to this puzzle are now clicking together.

According to experts the mind control, parts of Microwave Weapons are most success-ful on people who are taking mood altering drugs, like anti-depressants. In 2008, news reports stated that these sorts of drugs were being found in around 24 major public water supplies in the USA. And I have reason to believe it is not from "run off."

The general public being drugged through our public water supplies suggests that the radio wave brainwashings are aimed at most of the USA population.

Around the late 1980s or early 1990s I remember reading a medical report about the new development of technologies that would enable laser surgeries to be performed by satellite, on a patient, while he or she is in their own home. This appears to be hushed since then. Why? It appears that these weapons are being used on Targeted Individuals. And I know a woman who appears to have been used for remote technological medical experimentation.

I believe that this truly is a lethal Technological Holocaust, of an unmeasurable scope, which needs to be exposed and stopped for the safety of all of humanity. And it seems like the only way to stop it is if enough people find the Heart to realize what is happening, and then find the Courage to openly stand up against these crimes. (I feel that radio wave blockers should be immediately legalized for common citizens.)

I hope the media runs with this soon, because the psychological Freedom and health of media and government personnel has been as much at risk as ours. . .perhaps even more so.

Return the Constitution

*There are microwave weapons aimed at them and me
And people dying for telling us, but we're too blind to see.
There's an evil darkness aiming for control
And we just accept it, because we don't know.
Eugenics didn't stop with the Hitler we degrade
But the Truth is being buried deep inside our graves.
The freedom that we boast of in the "good old USA"
Has been being secretly taken away.
Lets take it back - Lets take it back
Return the Constitution to it's original track.*

This site also hopes halt the growth of deceptive recruiting of unaware citizens into a covert program that is used to target fellow human beings.

I feel that the secrecy surrounding covert technological crimes is enabling their contin-uation and growth, which is a serious threat to the future safety of all of humanity.

Many may wonder why the government secrecy around this continues. The answer to this question, and the existing technological crisis, is NOT to blame or go against our government or media, because they probably have good reason to have not informed the public. Perhaps they are concerned that the exposure will cause a panic. . .or perhaps they want to wait until they have the problem more under control. I don't want to judge them, although there are times when I have. I want to trust that they are obviously doing what they think is best. And I need to also follow my own heart and do what I feel is right.

A part that now deeply concerns me is the fact that I know only three families of people who work for the FBI and all three of those families appear to be being targeted in various ways. Since I realized this, the question that haunts me is "if the FBI is also being targeted, who is going to help the rest of us?" But I am working on letting go of this concern and trusting that there are some good people in our government, who are working on this problem.

But I hope it is exposed soon, because I do not feel that the public will panic if the exposure is handled properly - with honesty and reassurance and compassion and encouragement for people to pull together and help each other instead of overwhelming government agencies with phone calls.

I feel that most people can sense that there is something wrong and that not understanding exactly what it is has them in a state of confusion. I feel that the truth coming out can reassure many, validate many and give us all an opportunity to pull together and help each other through it.

Though I am being more heavily targeted, I feel that I must continue helping to expose these crimes, because the magnitude of pain and suffering that I have experienced, and witnessed in other victims, is just too horrible to remain silent about. . .especially since I see no promise of its end. I pray that this site helps to instigate the end of the technological targeting of humanity.

SILENCE HURTS!

My soul cries from its tortured depths
For the freedom we've not attained yet.
I hope you open your heart and see.
This covert war is not meant to be.

PLEASE LET YOUR HEART REALIZE WHAT IS HAPPENING AND LET YOUR COURAGE STAND UP AND HELP REGAIN OUR FREEDOM

If You Look With Your Heart You Will See

"The ultimate tragedy is not the oppression and cruelty by the bad people, but the silence over that by the good" ~ Martin Luther King Jr.

Chapter Two

The Covert War

Its Worth Fighting For -

Humanity's Right to Freely Live and Love and Laugh and Cry.

Its Time to Fight Like Heaven!

I am still deeply concerned that the covert "rescues" are actually enslavement and that technological modes of "protection" could be a sly enslavement performed by those who may also be placing filters in detection technologies, in order to prevent detection of the low frequencies that are used for mind control. Help spread the word on this.

The Covert War

A Covert War Thrives in the Shadows of a Technological Holocaust

There is a growing concern about this covert war and its effects on the vast unaware population. (I hope there is, anyway! ;-)

I've struggled to understand exactly how this began, how it has grown to such a degree, and how it operates. I can only go by my own experiences and instincts. But it appears that there may have been a covert program set up to help fight terrorism and that a global covert war has been secretly raging. Most citizens seem to not be aware of it, although many are being used in it and some of us have been being either destroyed by or enslaved into the dark side of it, which utilizes mind control technologies and sly "protection" that really enslaves people.

In the 1970s President John F. Kennedy had warned of the dangers of handling this problem covertly but then was murdered, probably due to his openness about it, which may have forced other government officials into handling it covertly out of concern for their own safety. Below is a video where JFK addressed this critical issue in the 1970s. He openly warned of a covert organization that is a serious threat to humanity. When I first watched it I felt validated - his description of their tactics perfectly mimic that of those who are still targeting humanity. I feel that it is the same entity.

Perhaps now is a time when JFK's wisdom can be of the greatest value to humanity;

*"I want to talk about our common responsibilities in the face of a danger. . . the dimensions of its threat have loomed large on the horizon for many years. Whatever our hopes may be for the future - for reducing this threat or living with it - there is no escaping either the gravity or the totality of its challenge to our survival and to our security - a challenge that confronts us in unaccustomed ways in every sphere of human activity. . . the very word secrecy is repugnant in a free and open society; and we are, as a people, inherently and historically opposed to secret societies, to secret oaths and secret proceedings. We decided long ago that the dangers of excessive and unwarranted conceal-ment of pertinent facts far outweighed the dangers which are cited to justify it. Even today, there is little value in opposing the threat of a closed society by imitating its arbitrary restrictions. **Even today, there is little value in insuring the survival of our nation if our traditions do not survive with it.** . . And there is very grave danger that an announced need for increased security will be seized upon by those anxious to expand its meaning to the very limits of official censorship and con-cealment. That I do not intend to permit. . . Today no war has been declared, and however fierce the struggle may be, it may never be declared in the traditional fashion. Our way of life is under attack. Those who make themselves our enemy are advancing around the globe. The survival of our friends is in danger. . .no war ever posed a greater threat to our security. . .for we are opposed, around the world, by a monolithic and ruthless conspiracy that relies primarily on covert means for expanding its sphere of influence - on infiltration instead of invasion, in subversion instead of elections, on intimida-tion, instead of free choice, on guerrillas by night, instead of armies by day. . .its preparations are concealed, not published. Its mistakes are buried, not headlined. Its dissenters are silenced, not praised. . . This is a time of peace and peril, which knows no precedent in history. It is the unprece-dented nature of this challenge that also gives rise to your second obligation, an obligation which I share. And that is our obligation to inform and alert the American people - to make certain that they possess all the facts that they need, and understand them as well. . . **I have complete confidence in the response and dedication of our citizens whenever they are fully informed. . . it is to the printing press - to the recorder of mans deeds, the keeper of his conscience. . . that we look for strength and assistance, confident that with YOUR help, man will be what he was born to be - free and independent."***

JFK Secret Societies Speech (full version)
https://www.youtube.com/watch?v=zdMbmdFOvTs

PLEASE Help Break the Silence

I am one of the victims who have been being slowly destroyed. I have witnessed dozens of other good decent people also be targeted. And I am working hard to be as objective as I can. I do not want to blame the USA government, like many do, because I feel that the good parts of our governemnt have been doing what they feel is right to retain our Freedom. . .and this is honorable, no matter what the outcome. However, all of my instincts - my heart and soul are strongly feeling that the Covert War must now stop, in order to regain our Freedom. I am deeply concerned that too many people are being recruited or enslaved into the dark covert program due to a lack of awareness, "protection" that slyly enslaves and freely used mind control technologies . . .and that victims who refuse to join are being tortured...etc. I strongly feel that it ALL needs to be exposed and stopped in order to regain our Freedom.

PLEASE HELP STOP THE COVERT WAR

Gulf War Syndrome = Microwave Targeting?

Is it a coincidence that the "Gulf War Syndrome" Symptoms mimic that of microwave targeting?

As I sat in a service station, waiting for my car to be checked for inspection, I picked up a magazine and found that the Sept 2014 issue of "Discover Magazine" contained an article by Florence Williams entitled, "An Invisible Enemy," which explains that "between 170,000 - 250,000 veterans," who served in the Gulf War, experienced unexplainable symptoms, which included head aches, joint pain, memory problems, extreme fatigue, difficulty focusing, gastric distress and respiratory disorders. An article in "USA Today" added in the symptoms of "sun sensitivity," "neurological problems" and "swollen bleeding gums"...etc.

Williams stated that among the veterans who were injured was Carolyn Kroot, and that, "Like many other Gulf war veterans with similar symptoms, Kroot was told that her problems were due to mental stress and best treated with an anti-depressant."

Even after being told that she suffered from a mental disorder, and after it was proven to be brain damage twenty years later, Carolyn still yearns to understand the true cause. The article wrote that Carolyn said, "There is something here, a reason why these things are wrong with my body," she says, "I have some hope that there may be a way to treat it someday." This deeply touched my heart and raised pains of my own suffering and hoping.

I wonder if Gulf War Syndrome is really radiation illness caused by microwave targeting, because I've experienced YEARS of the same types of symptoms. I also had two mild seizures by the year 2002. From around 1999 through 2005 I had spent thousands of dollars on medical tests that showed nothing wrong with me and this was worse than frustrating. The worst of my obvious physical symptoms consistently existed through 2005 and 2006 and also included body bloating, unusual hair loss and episodes of dizziness and ringing in my ears. There was obviously something wrong, but no one could or would tell me what it was.

I deeply related to Carolyn's yearn for the truth that can offer a solution - a possible end to the suffering. Knowing that something is horribly wrong, and not understanding what it is can be more disconcerting than the actual symptoms, because it leaves us with no possible way to even try to correct the problem - it leaves us feeling trapped in a mystery that offers no avenue for treatment and no possibility of recovery.

In 2006 I was told, by the people who target me, that I had lupus, but I kept feeling that something else was wrong - that there was more to it than that. During an episode of OBVIOUS neurological problems a doctor had tried to call it depression and gave me a prescription for antidepressants, which I never took, because I knew it was not that. (There is something horribly wrong with the push for antidepressants, even for obvious physical illness!)

It wasn't until 2011 when I finally began realizing that I am a victim of microwave targeting. Some levels of the radiation suddenly stopped shortly after I figured it out. . . probably to hide the crime. Although the targeting then vamped up in other ways, finding out the truth was a HUGE relief for me, because knowing what I am up against opened doors to things I can do about it. I now know what to seek help for, although help is not yet here for victims of microwave targeting. But because help and protection is not here for me I strive to bring public awareness to this crisis with the hope that it will help bring an end to criminal use of microwave weapons, which extends far beyond inflictions of the types of physical illness mentioned in this article. (My website for this purpose is www.targetedinamerica.com)

The possible microwave targeting of "between 170,000 - 250,000 veterans," raises more concern than that for their safety and destroyed lives, although this is certainly more than enough to trigger the tears that are welling in my heart. It is my understanding that these forms of microwave targeting/torturing are done in the brainwashing and recruiting process for the covert program, which appears to be performing a take over of the USA while targeting us.

Since I read the initial article, my heart has been filled with concern and questions. How many of the inflicted Gulf War veterans took the antidepressants, which merely aid the mind control process? How many refused to take them and are surrounded by families who are not being allowed to understand or care or support them? How many are now homeless and struggling to just survive? How many have been shoved into institutions under a false "mental illness" diagnosis? How many still suffer through tortures that no human being should have to continue experiencing in a world that already has the answers and cures? And how many have been forcefully or deceitfully recruited into covert operations that are destroying our freedom? Its horrible to even think that our veterans would be used as puppets against their own country and fellow citizens. But its not at all unrealistic. I have noticed MANY veterans involved in the covert targeting of me. And this is horribly sad for ALL of us.

I want to think that those who are in charge of informing the public have not been aware of the different types of microwave targeting. I feel that the good in our government has valid reasons for keeping it under wraps, if they were or are aware. Perhaps they do not realize how difficult it is for victims to deal with.

Whether the "Gulf War Syndrome" is caused by microwaves or not, I hope this article helps government officials to realize how difficult the mystery is for victims of microwave targeting. We suffer in ways that no human being should have to continue suffering. . .and that suffering extends into the hearts of our loved ones who are not being allowed a chance to understand what is happening to us. We desperately need all levels of the targeting to be exposed and stopped.

If the Gulf War Syndrome is really microwave targeting, (and I feel that it probably is) it could offer hope for other Targeted Individuals. The article in "USA Today" said, "Using fMRI machines, the Georgetown University researchers were able to" find physical brain damage. Do certain types of microwave targeting cause testable brain damage? According to Tim (Turan) Rifat the microwaving causes certain types of cell structure damage that can be detected with proper medical testing. But the big question is. . .when will anyone honestly test us and protect us from further harm?

"The ultimate tragedy is not the oppression and cruelty by the bad people, but the silence over that by the good" ~ Martin Luther King Jr.

I am still deeply concerned that the covert "rescues" are actually enslavement and that technological modes of "protection" could be a sly enslavement performed by those who may also be placing filters in detection technologies, in order to prevent detection of the low frequencies that are used for mind control. Help spread the word on this.

Lupus Cause and Cure?

The medical profession has been baffled by this mysterious illness, who's symptoms come and go with no apparent cause, had suddenly appeared just a few decades ago, and was originally found primarily in Native American women.

Is Lupus microwave induced? I believe so. Do the original inflictions of Lupus, being primarily on Native American women point to eugenics based targetings? I believe so, because of my own personal experiences and the statistics I had found in 2006. It could also be due to microwave weapon experimentation, but then why just the women and not also the men? Eugenics is said to focus more on weeding out women, because we are the child bearers.

As I struggled to survive Lupus, I was forced to realize things that have been difficult for me to fully face - things that may also be hard for you to swallow, but please read and share this with an open mind and heart for the sake of those whom it may help.

In 2006, I began foolishly thinking that Lupus was being caused by a negative energy, which was being intentionally directed at me through a dark occult that was targeting me. Sounds crazy, right? There were times when I thought so too, although my insights and dreams were affirming this belief. Because of my spiritual focus, and my technological ignorance, I had assumed that it was being delivered via some sort of spiritual/energetic method, although it seemed far too strong for that.

In 2011, a private investigator opened a door that eventually lead to my realizing that I was a victim of a lethal targeting program and pieces to an ugly puzzle began clicking together.

My initial thoughts about lupus were actually not very far off. It appears that the targeting is being done with microwave/radio wave technologies, which (depending upon which frequency is used and where it is directed) is reported to cause leukemia and a large variety of other illnesses. And I firmly believe that the microwave targeting also causes Lupus and that these weapons are being used in a covert eugenics movement, as well as on Primary Targeted Individuals.

I understand how unbelievable this sounds. It's still a shock to me, although I have directly experienced and witnessed enough of it to feel sure of this.

In 2006, after I began publicly stating that Lupus was caused by energy being directed at me, the targeting vamped up and I started experiencing direct attempts on my life in ways that would make it appear like an accident. Those who target me began trying to distract me and brainwash me into forgetting

29

about it. They suddenly stopped giving me Lupus (or have altered new medical tests) and began trying to convince people that I never really had it - that I was making it up. (Aims to discredit me have hit multiple levels.) But they didn't realize that I had written documentation from the doctors who ran the initial tests. I do still get targeted with microwaves, but in different ways.

I've been being targeted in ways that most people would not believe could happen in the USA or any other Free country. Every day of my life has become a fight to survive and expose these crimes.

Now that you've read this, the cure for Lupus should be obvious: Humanity needs to rise into a peaceful, but forceful, fight to expose criminal use of radio wave technologies (microwave weapons). It may also help victims to listen very closely to their instincts when deciding whether or not to use modern medicine for lupus. In 2007 I refused medical treatment and found the ORIGINAL Essiac formula very helpful. A lot of clear spring water and exercise also helps. Check out www.healthfreedom.info

Please help pray for humanity to be protected from lethal technological targeting, which may be harming growing numbers of people with far more than just Lupus. According to the experts, radio wave technologies can remotely inflict leukemia, diabetes, various forms of cancer, a variety of lung problems, heart attacks, behavior modifications...etc.

Please help spread the word about this,
Even if you don't completely believe it,
Because it could save people's lives.

P.S. It appears that statistics on the web have been altered since my initial 2006 investigations. The web appears to be being filled with intentional misinformation about microwave technologies and their long term use on humanity. Blogs I have put information on are being tampered with and my computers, phones and emails have often been remotely invaded. The web is not safe, private or accurate when it comes to these covert crimes against humanity. Please listen to the Heart of your own instincts above all else.

It is more than likely that when people hit the severe levels of Lupus, what they are feeling is REALLY happening to them. . .THAT THEY ARE BEING ATTACKED BY CRIMINALS WHO UTILIZE TECHNOLOGIES, WHICH PROJECT IMAGES, VOICES. These are purely sadistic crimes. "V2K" is what the military calls the visual and audio part of this process. And the process of misdiagnosing victims of technological abuse seems more cruel than the attacks, because victims would have a sense that something is wrong, but then being told that it isn't also pushes them into distrusting their own instincts.

PLEASE LISTEN TO YOUR OWN HEARTS AND INSTINCTS. . .AND STRIVE TO HELP FIND A REMEDY FOR THESE HORRIFIC CRIMES AGAINST HUMANITY.

Eugenics - Population Control

Did the horrors end with Hitler or did they continue in more covert ways? The Truth may never fully step out into the open, but when we listen to our instincts and our own Hearts we can feel it.

In a book called, "THE MEN BEHIND HITLER - A German warning to the world," Bernhard Schreiber exposed the field of psychiatry as being connected to an ongoing holocaust. He wrote, "Hitler was an evil man and no one would want to assert that he was not responsible for the things that happened in Germany, but in blaming Hitler for all the evils one is overlooking a considerable number of those who are truly responsible, people who are being allowed to pursue their course to similar ends all over again - nothing to stop them."

In another part of his book, which appears to have been erased, Schreiber had spoken of Dr. Paul Joseph Goebbels bragging, in 1944 - after Hitler's Reich had fallen, about a secret weapon - a "death ray" that would lead to the "rebirth of the Reich." The fact that this statement was erased has great significance, because it shows us what is being hidden.

However Schreiber's conclusion, where he tried to figure out what the weapon was, remains at the end of his book. He wrote, "The Nazis may have been disbanded, but the psychiatrists still linger on among us. Maybe this is the secret weapon Goebbels boasted about which would lead to the rebirth of the Reich - not a super-bomb and not a death ray, but a blueprint for a psychiatric slave state."

My instincts are telling me that this mystery weapon, which Geobbels boasted of in 1944, is microwave weapons that are capable of inflicting a "psychiatric slave state" through shooting beams of radio waves into the brains of human beings, especially when pharmaceuticals are added to the equation. . .as I have been saying all along. Microwave weapons are also capable of inconspicuous murder. According to British General Barrie Trower, microwave weapons can remotely inflict terminal illnesses like leukemia, diabetes, cancer...etc., at the flick of a switch. And I have believed, since 2006, that Lupus is caused by energy being directed at people. Again. . . I now believe that that energy is microwaves.

A Microwave weapon massacre can remain virtually undetectable in a world where the general public is not aware of the existence of microwave weapons and are not allowed to possess unfiltered microwave detection technologies.

Do the original inflictions of Lupus, being primarily in Native American women point to eugenics based targetings? I believe so.   Is it true that microwaves can be used to damage part of our brains and literally turn us into numb puppets that do our master's bidding? I believe so, due to witnessing it happening to peo-

31

ple. Is this being done by the same people who are doing the eugenics based targetings? Most likely. Are those of us, whom they perceive as less worthy, being either killed or turned into brain damaged slaves? It appears so. And this, along with other forms of remotely inflicted mind control, may be the most dangerous thing that humanity will ever have to face and stop.

I'm no expert on this, but I feel the overall picture of what is happening and that it is criminal and can not be allowed to continue, in order to preserve the Freedom and psychological safety of ALL of humanity. (My guess is that the underground Reich is not the only group that utilizes microwave weapons.)

Beyond the lethal technologies, that are being used on human beings, some reports state that substances, which are KNOWN to be unhealthy, are being put in food, seeds, water, vaccines and pharmaceuticals. Are these also subtle efforts to cut back on the earth's population?

There has been a strong push for people to not use herbal remedies. Rene Cassie - the nurse who successfully experimented with a Native America cure for cancer, was faced with all sorts of problems from agencies that apparently did not want people to have a cure for cancer and diabetes. Perhaps some of the problem is from those who are either involved in eugenics based targetings of human beings and those who are tied up in preferring to make money off of our illnesses instead of letting us have simple natural cures.

I have also read reports which state that illnesses like aids and other new ailments, that just suddenly pop up out of no where, are man made in intentionally inflicted. (Lupus is in this category.)

In the broader scheme of things. . .are weather modification technologies being used to cut back on the Earth's population, through what has been perceived as "natural disasters." And the same question should stand for laser weapons that can cause earth quakes and possibly even rupture volcanoes...etc.

I think that the most dangerous methods of harming and/or inconspicuously murdering human beings are microwave weapons, because it is a slow torturous death that leaves a victim confused with the process of sensing that something is horribly wrong. . .but due a lack of awareness, of criminal use of microwave weapons, the puzzle remains unsolved and the victim dies without ever having the chance to come to terms with what is happening to them.

Sometimes my anger at the thought of these types of crimes continuing makes me want to scream; "SERIOUSLY!??! Aiming for complete control of human beings through mind control technologies!!! Slowly murdering people with microwave weapons!!! Murdering people through creating floods and earthquakes...etc.!!! Harming people through placing toxic materials in water, food, vaccines or pharaceuticals!!! Torturing or killing anyone who is a threat to the secrecy and success of such madness!!! How on EARTH can this be happening in America and other free countries???!!! "WE ARE HUMAN BEINGS WHO

DESERVE KINDNESS AND RESPECT AND THE RIGHT TO LIVE AND WORK AND LOVE AND FEEL AND THINK FOR OURSELVES!!!!"

Humane methods of population control would be things like. . . strongly encouraging people to NOT have children and OFFERING sterility as an OPTION. I fully believe that if the public were properly informed on the dangers of over population, MOST citizens would do what they can for the preservation of our Earth and the health of humanity.

If just a fraction of the moneys used to murder us were put into a drive to inform the public and encourage CHOICES, the same results can happen without the crimes. But would that be too humane for those who are now in control? It appears so.

The question now is, "What can we do to turn this around? What can we do to end this holocaust and regain our basic human rights?" What can you do?

PLEASE LISTEN TO YOUR OWN HEARTS AND INSTINCTS AND STRIVE TO HELP FIND A PEACEFUL REMEDY FOR THESE CRIMES AGAINST HUMANITY.

Weather Modification

Mother Nature fills her place
Far better then the human race.

It appears that weather modification technologies have been utilized through the past several decades. Although most civilians are not aware of their existence, they are reported to be fully operational and can be used to redirect storms, stall storms, create storms, diminish storms...etc. The effects can be good or bad, depending on the intentions of the users.

International concerns in the 1970s proves that these technologies do indeed exist. And common sense says that they were invented to be used. Technologies like the Russian SURA and the American HAARP are reported to be among those used for weather modifications. And there are sure to be others scattered around the globe. A variety of scientists, who are/were monitoring the use of radio waves (microwaves) for weather modification have reported these technologies probably being the true cause of global warming. I've read reports stating that some of these technologies can also be utilized as laser weapons, which can trigger earth quakes.

I'm no expert on this subject, but I believe those who are. My common sense realizes that there is something unusual happening with the weather, and what is being called "natural disasters," through the past few decades. My insights have shown me that the Alstead, NH flood was instigated by weather modification technologies stalling a storm above an area where a culvert is said to have been plugged. And my insights are also telling me that these technologies have been used to create other disasters and that there may be more to come if these crimes are not exposed and stopped. So spread the word quickly, please.

Weather Wars Documentary Promo Scott Stevens
https://www.youtube.com/watch?v=gZwh8WLTMUU

Chapter Three

Mind Control Steals Hope

Its Worth Fighting For -

Humanity's Right to Freely Live and Love and Laugh and Cry.

Its Time to Fight Like Heaven!

Mind Control Steals Hope

What will become of humankind if not allowed a free mind?

This is not a "theory." Its a fight for our lives. Its not a matter of if you "believe it" or not - its a matter of if you are aware and if you can care to help restore our safety and freedom.

Most people are aware that radio waves can be shot down, from satellites, for the purpose of internet access. But most people are not aware that various frequencies of encoded radio waves can also be shot into whole communities or directly into the brains of individuals, in order to perform behavior modification - mind control. And this is happening to people.

But it all sounds so much like science fiction, and is so outrageously inhumane, that the natural response is to slip into shocked denial, overwhelm, or blind disbelief, but there is not time for these types of responses. This situation needs humanity's awareness and attention. If these crimes are allowed to continue and grow, ALL of humanity could be effected. Too many probably already are. NOW is the time to take steps toward insuring humanity's future safety and freedom.

I'm no expert on mind control technologies. But due to my experiences, things that I've witnessed and my insights, I feel certain of their use on people since at least the 1970s. . .and of the destruction that this is causing, not only to individuals, but also to families, communities, countries...etc.

As I look into the past of victims whom I know, and face this devastating reality, I feel literally terrified for humanity. Even just mild levels of long term technological mind control prevents the natural process of personal and spiritual growth. On massive scales, this could literally destroy the Heart and Soul of humanity. And I feel that this is already happening. I feel that technological mind control is at the core of most of the problems we are faced with in our world today. This is truly the most dangerous thing humanity has ever been faced with and it must be stopped as quickly as possible. So, PLEASE realize what is happening, help break the lethal silence that enables its continuation, and do all that you can to help bring it to a PEACEFUL end.

If you are doubting the reality of technological mind control I beg you to break through the walls of doubt and let your Heart realize that there TRULY is the existence of "behavior modification technologies" - Nicola Tesla had experimented on the effects of radio waves being remotely shot into to the human brain and body as early as the 1930s. And it should not be much of a stretch for anyone to realize what could happen if technologies, which can effect human behavior, fall into the wrong hands - hands that either have no understanding of the natural process of personal growth or hands that have no respect for human rights or Freedom. And I feel that this is exactly what has happened.

Please read my article on the Tesla Technologies;
www.targetedinamerica.com/atesla.htm

According to reports I've read, radio waves (microwaves) can be encoded with subliminal messaging or be set at frequencies that interfere with our brain's natural rhythms and functions. Results appear to depend on which part of the brain is intruded upon, what frequency and intensity is used, and can range from mild brainwashings, mental numbness and interrupted thoughts. . .to brain damage and complete technological control of a human being.

According to experts technological mind control is most successful on people who are taking mood altering drugs, like anti-depressants. I have witnessed the Truth in this and strongly feel that there is a direct link between this and the push of these sorts of pharmaceuticals in medical fields. Is it a coincidence that, in 2008, news reports stated that these sorts of drugs were also being found in around 24 major public water supplies in the USA? I don't think so.

Drugs found in about 24 public water supplies
https://www.youtube.com/watch?v=LYD-qf9pZAg

In 2013 a secretary at a New Hampshire Environmental Protection Agency told me that these drugs are STILL being found in our public water supplies! Reports state that it's from the "run off" of pharmaceuticals that are improperly disposed of. But those who are coming to this conclusion are probably not aware of the microwave and pharmaceutical targeting of humanity.

Please read my article on "Connecting the Dots Between False "Depression" and other "Mental Illness" labels, Microwave Targeting, Harmful Psychiatric Pharmaceuticals and Eugenics" www.targetedinamerica.com/psychiatry.html.

I feel that the general numbness, the growing lack of care and growing levels of discord in the world are mostly due to criminal use of radio wave technologies and their aiding psychiatric pharmaceuticals. I also feel that technological mind control plays a HUGE role in the recruiting/enslavement of citizens into the covert program that is used to help target people. The criminals who lead this program appear to be sadistically aiming to cause chaos, pain and suffering. The general rise in crimes have been being blamed on economic difficulties. But the truth is that, by nature, human beings tend to pull together during tough times - letting our Hearts help each other through. And the fact that the opposite appears to be happening has me deeply concerned. I feel that. . .

What is now happening is not natural, not normal,
NOT "meant to be" and is a criminal assault on an
unsuspecting and defenseless populous!

Technological and pharmaceutical mind control prevents the natural process of personal and spiritual growth and blocks our Hearts. The harm that this inflicts cannot be over-stated, because it hurts us on EVERY SINGLE level of our existence, especially for long term victims. Most victims seem to be completely unaware of what is happening to them. But some, especially those who are sensitive to radio waves, or those who are being more heavily targeted in other ways, are aware and have been trying to expose and report it through the past few decades. But too few have been listening. Please listen - please at least give this the benefit of your doubt.

If you research this (and I hope you do) you will need to listen closely to the Heart of your own instincts above all else. There is a lot of misinformation on the web. But hopefully the links I supply here will help you to realize that these technologies do indeed exist.

In 1976 Time Magazine said, "Last month the U.S. confirmed that for some 15 years the Soviet Union has been beaming microwaves at the hulking nine-story U.S. Embassy..." Find more on: http://www.time.com/time/magazine/article/0,9171,911755,00.html

Quote by (the late) veteran author Jim Keith: "Brain-computer radio communication has long been considered impossible by the majority of people and has consequently been relegated to science-fiction, but the fact is that the technology had been developed into reality by at least the 1960s, during which time the initial experiments were being performed on unwitting subjects." www.whale.to/b/keith.html

What are these technologies and how do they work?

Again, I am no expert on these technologies, but I trust my experiences and the reports of those who are valid experts. Technologies like the Russian SURA, the American HAARP, Gwen Towers and Satellites, are the primary suspects in technological mind control part of the targeting. Many appear to be blaming only cell phones and cell towers. Some believe it is from ground based microwave weapons that can be set up near a victim's residence. Common sense knows that ANY technology, which can emit and direct radio waves (microwaves), can be criminally used.

Due to what I've witnessed and directly experienced, I feel certain that at least some of the technological targeting is being done VERY remotely. I've been being targeted with microwave and laser weapons, which appear to have the capability of shooting beams of radio waves FROM THE SKY and quickly circling around and coming at me from different directions and angles. I've experienced this when I've been in areas where there are no houses, vehicles, cell towers...etc. And there have been times when the beams of radio waves have left circular melted spots on my cold windows in the winter, which indicate that they are coming from the direction of the sky. The only technologies, that I know of, which can do these sorts of maneuvers from the sky, are satellites and

Microwave weapons - the types of technologies that bounce radio waves off of the ionosphere, in order to be redirected.

But no matter which technologies are being used and how it is being done it needs to be stopped. I feel that this is the most dangerous thing humanity has ever had to face and stop. . .and that this MUST be done quickly.

Interview with Microwave Expert General Barrie Trower
https://www.youtube.com/watch?v=kvn-8ITy0oc

Find more information on radio wave technologies on
www.targetedinamerica.com/weapons.htm

Quote by President John F. Kennedy, in his 1961 "President and the Press" speech, which addressed the covert targeting and its secret societies; "Today no war has been declared - and however fierce the struggle may be, it may never be declared in the traditional fashion. Our way of life is under attack... we are opposed, around the world, by a monolithic and ruthless conspiracy that relies primarily on covert means." "there is little value in opposing the threat of a closed society by imitating its arbitrary restrictions. . .I want to talk about our common responsibilities in the face of a danger. . . the dimensions of its threat have loomed large on the horizon for many years. Whatever our hopes may be for the future - for reducing this threat or living with it - there is no escaping either the gravity or the totality of its challenge to our survival and to our security - a challenge that confronts us in unaccustomed ways in every sphere of human activity... This is a time of peace and peril, which knows no precedent in history. It is the unprecedented nature of this challenge that also gives rise to... our obligation to inform and alert the American people - to make certain that they possess all the facts that they need, and understand them as well... I have complete confidence in the response and dedication of our citizens whenever they are fully informed."

The time to heed JFK's Wisdom feels long over-due.
https://www.youtube.com/watch?v=zdMbmdFOvTs

Who is Doing the Targeting?

I have tried hard to figure out exactly who is responsible so that it can be stopped. But I have been being too heavily targeted to get very far with this part. However, there are a few obvious things. . .although, the targeting is so manipulative that almost nothing is as it seems - the obvious conclusion is not always the true answer. Please remember this.

According to a dream I had, those who target humanity are dark/criminal people who are working underground. I fully trust this, especially since having and sharing the dream brought on heavy rounds of targeting, which included an attempt to harm me, deaths of loved ones, a suspicious fire that raged through my home...etc.

39

Many blame "the government." Obviously governments around the globe have been experimenting with and utilizing radio wave technologies for various things, especially for military purposes. This is evident in the links I have on this page. But it is also a matter of documented history that governments, as well as the medical field, were interested in things like Nicola Tesla's inventions. It is safe to assume that many organizations have openly and/or secretly experimented with technological mind control. But how many of these organizations would use it to intentionally inflict harm on and/or control harmless citizens?

Due to the cruel nature of various parts of the technological targeting of human beings, which I have experienced and witnessed, there seems to be some sort of dark occult leading it. Much of the targeting appears to have an aim to take over America (probably also other countries) and ultimately. . .all of humanity. Whoever would aim for this sort of control, and have such little regard for human health and human rights, is surely akin to Hitler and those who supported him.

I feel that they have been targeting BOTH government officials and common citizens. One of their patterns is to manipulate things so that the innocent (their victims) get blamed for what they do. In this process they pit people against each other - pit family members against each other - pit governments against citizens and citizens against governments, pit countries against each other...etc. They seem to enjoy instigating wars. And it appears that this is exactly what they are doing. With the aid of pharmaceutical and technological mind control, and forced or deceitful recruitings, the covert war rages into their success.

I feel that it is up to us to not allow their manipulations to gain anymore success - we are all in this together. . .and together we can rise up to stop it.

Please help break the lethal silence that also enables their success.

I pray that leaders of nations, around the globe, unite to regain humanity's Freedom - unite into a public stand that helps citizens to understand what is happening. . .so the HEARTS of citizens, including government officials, can unite into a PEACEFUL stand to regain our freedom. . .supporting each other through that process.

Our world has been being secretly torn apart
Through breeches of mind and blocks of heart.
But Light will reign as pure as snow
To Save us from the final blow.
Please help it to.

The following quotes and links are to merely prove that these technologies do indeed exist. Research has revealed only that which has been publicly exposed. I feel that there is a lot that has not been exposed, both in the arena of the technologies and the organizations who exercise criminal use of them.

A quote from the Space Preservation Bill proposed by Congressman Dennis J. Kucinich in the U.S. House of Representatives in 2001. This was a Bill "*To preserve the cooperative, peaceful uses of space for the benefit of all humankind by permanently prohibiting the basing of weapons in space by the United States, and to require the President to take action to adopt and implement a world treaty banning space-based weapons.*" This was an aim to prevent harm "*through the use of land-based, sea-based, or space-based systems using radiation, electromagnetic, psychotronic, sonic, laser, or other energies directed at individual persons or targeted populations for the purpose of information war, mood management, or mind control of such persons or populations.*" **http://thomas.loc.gov/cgi-bin/query/r?c107:chemtrails**

Quote by Philip Coppens in "The Russian Woodpecker: Experiments in Global Mind Control?"; "In April 1953, CIA Allen Dulles gave a lecture at Princeton University, detailing Soviet developments in the field of mind control. He stated **they were out to control the mind of free men, both individually and collectively. . .** Dulles proclaimed that the Cold War was moving into a new era of psychological warfare, which Dulles characterized as the battle for men's minds. *"We might call it in its new form brain warfare. . .In the 1970s, some of this "secret war for our mind" was exposed. . ."* Find more on my "Articles" page or on this link: http://www.philipcoppens.com/woodpecker.html

Quote by British Psychoanalyst, Carole Smith in her article "On the Need for New Criteria of Diagnosis of Psychosis in the Light of Mind Invasive Technology"; "*We have failed to comprehend that the result of the technology that originated in the years of the arms race between the Soviet Union and the West, has resulted in using satellite technology not only for surveillance and communication systems but also to lock on to human beings, manipulating brain frequencies by directing laser beams, neural-particle beams, electro-magnetic radiation, sonar waves, radiofrequency radiation...*" http://www.globalresearch.ca/index.php?context=va&aid=7123

Gordon Duff - senior editor of Veterans Today stood up for Targeted Individual's mind control claims; "At one time, we thought of all of them as 'tin foil hat' conspiracy theorists. This was until we were able to break through the encoding with some mobile communications devices. . . The concerns go well past simply a few thousand targeted 'trouble makers'..."

Quotes by expert Turan (Tim) Rifat: "Russian and American research has found that pulse modulated microwaves (as used for mobile phones) can, when modulated with ELF which mimics specific brain patterns, change the behavior of the victim with the flick of a switch."

   "Precisely modulated microwave radiation is used to influence brain function. Human behavior and reactions can be entirely controlled by using pulse modulated microwave EM radiation" There is "a catalogue of every specific brain frequency for each mood, action and thought. . .there is one for anger, suicide, hysteria, trauma, serial killing, paranoia, lust...etc."
http://www.whale.to/b/rifat.html

Komsomolskaya Pravda stated that, " In 1924, chairman of the academician

41

council of the Animal Psychology Laboratory, brilliant animal trainer Vladimir Durov wrote a book on animal training and told about his experiments on hypnosis applied to animals. In 1932, the Bekhterev Institute of Brain named after the scientist was officially charged to conduct experiments on distant interaction. In 1965-1968, the Institute of Automatics and Electroenergetics based in Novosibirsk studied mental communication between humans and animals. The materials of the study were classified and were never published officially."
http://english.pravda.ru/science/tech/14-08-2007/95965-psychotronic_weapon-0

It would be foolish to assume that these technologies have not also been being used on human beings within the same time frames. (My experiences say this has been done since the mid 1970s, but it appears to have begun much sooner than this.)

Symptoms of Microwave Mind Control

Symptoms may slowly increase, come and go, or continue almost constantly. It seems to depend on how we are being targeted and for what purpose. Most victims appear to experience unnoticeable brainwashing and are not aware of what is happening. The most common noticeable symptoms are mental numbness and fatigue, a faint high pitched ring in ears, unusual neck tension and headaches and unusual thoughts that do not match feelings.

The most dangerous long term, general effect is harmful interference with our natural process of personal and spiritual growth - the destruction of our Hearts and Souls - the loss of psychological and spiritual freedom.

Long term mind control victims can become disconnected from their hearts and natural instincts and can seem narcissistic, which is one of the worse outcomes. The interference with our processes of feeling, thinking and sensing is LETHAL to our psychological health. We cannot grow and evolve and become all that we were meant to be while our brains are being intruded upon and manipulated.

Long term Heavily Targeted Individuals can also have symptoms like unusual tooth decay, unusual dry scalp, ridges on finger nails, chains of unusual medical inflictions, unusual discord between family members, unusual stress, mental confusion, forgetfulness, unusual or repetitive thoughts and mood swings, periods of diminished hearing and eye sight, sudden onset of unexplainable neurological problems and sharp pains shooting into head. . .as well as symptoms that mimic mental illness. . .primarily "depression," "split personality disorder" and "paranoid schizophrenia." Parts of the targeting seem sadistic/satanic with intentions to inflict psychological suffering and chaos. In severe cases, it appears that unaware mind control victims can be completely controlled by criminals who force them into suicide or things like lethal shootings at schools or navy yards.

Some people seem less susceptible to being completely controlled - perhaps those of us who are more creative and independent. People who are taking mood altering drugs, like anti-depressants, are VERY susceptible to being completely controlled.

Our own hearts and natural instincts can override SOME of the manipulations, if we are aware and are not too heavily targeted. BUT if we are not AWARE of what is happening we can easily confuse the intrusions with our own thoughts, feelings and instincts, and this can be devastating. . .to say the least. Through awareness and a strong will, we can avoid manipulations that are alien to our own nature when we are not too traumatized. But when our existing issues or feelings are triggered, or when we are also inflicted with heavy doses of emotional trauma, it is more difficult to resist.

Thus far, heavily targeted victims, who have become aware of what is happening, have not been able to obtain help or protection. We are being covertly tortured in the worst kind of prison - a prison that is filled with people who appear to be brainwashed into disbelief. Can you imagine?

PLEASE believe us. People need to become aware of what is happening and take HUGE peaceful steps toward bringing it to an end. THIS IS A SERIOUS CRISIS - Truly a Technological Holocaust! It is imperative that you listen to the heart of your own instincts on this subject, because your mind may not want to believe it and your heart may ache so much that it aims to slip into denial or overwhelm. . .but there is not time for these types of responses. This situation needs your heart to find the courage to face this and stand up against it.

**Please take COURAGE, wrap it around your HEART and rise
into a strong PEACEFUL stand to regain your FREEDOM.**

Take immediate action. Standing up in a strong and peaceful fight to stop criminal use of radio wave technologies. Write letters to the FBI, your Governor, Congressmen...etc. Tell them that you are concerned and ask them what they are doing to stop criminal use of radio wave technologies.

We will also be far better off when we, more fully, open and listen to our Hearts above what is being projected into our minds. . .and also do the following things: take as few pharmaceuticals as possible; drink only pure clean water; build or purchase radio wave detectors that do NOT have a filter, which prevents detection of the low frequencies used for mind control; Doing creative visualizations for spiritual protection can help dispel the negative brainwashings. (This can include imagining your body being filled with, and surrounded by, pure white light.) This may sound silly to some, but it REALLY helps; lead, tin and water can interfere with radio waves. Perhaps you can conjure up something which utilizes these substances. I feel that we must do whatever we can until our governments begin either strictly regulating or shutting down the technologies, which are being criminally used.

Note to medical, media and law enforcement personnel: PLEASE educate yourselves on microwave targeting and its effects and then investigate and help victims through performing the proper scientific tests for radio wave detection, cell structure damage, brain damage...etc., which can prove the targeting, instead of adding to our distress by assuming that we are "mentally ill," or prescribing psychiatric pharamceuticals, which actually help the technological mind control part of the targeting. (What if the lethal pill works only because the targeting backs off when it reaches its goal?)

43

Proving our sanity is an impossible task with anyone who does not realize what REALLY is happening and without the proper tests. We are already hurting too much. We need you to stand WITH us so that ALL of humanity can retain the freedom it needs, in order thrive and grow and evolve. Please research this.

There are many other links on the ones below. Please also do your own web searches, but be aware of misinformation on the web. Listen closely to the heart of your own instincts above all else, even with what I share here, because the writings of authentic people are sometimes invaded and altered by those who target us. If you look with your Heart you will find the Truth. And if you follow your Heart you will care to help bring an end to this holocaust. Please do.

Your mind's Freedom needs the Truth. Please search until you find it.

Among the labels of technological mind control are, Acoustic Psycho-correction, Synthetic Telepathy, Geophysical Weapons, Psychoneurological Weapons, Mind Machine, Psychological Language Machine, Acoustic Heterodyne Weapon, Electromagnetic Stalking, Microwave Weapons, psychological warfare, behavior modification technologies, Bio-communications Technologies, Radio Wave Mind Control, Microwave Mind Control, Bio-electromagnetic Technologies, Electronic Harassment, Directed Energy Weapons, Remote Neuro Monitoring, Brain Warfare, Bio-energetics, Electromagnetic (EMR) mind control, Electroenergetics, Geophysical weapons, psychoneurological weapons, M.I.N.D. - Magnetic Integrated Neuron Duplicator, Psychotronic weapons, Psychic Warfare...etc.

TECHNOLOGICAL MIND CONTROL IS PROBABLY THE MOST DANGEROUS THING THAT HUMANITY HAS EVER BEEN FACED WITH, BECAUSE IT PREVENTS THE NATURAL PROCESS OF PERSONAL AND SPIRITUAL GROWTH AND THE DANGERS OF THIS CAN NOT BE OVERSTATED.

CBS News Admits Navy Yard Shooter Possibly a Victim of Mind Control
https://www.youtube.com/watch?v=5u3EqLkKO_k

Science Channel Admits Microwaves Used in Mind Control
https://www.youtube.com/watch?v=zaPb3R5YTo4

CNN: Electromagnetic Mind Control Weapons (1 of 2)
https://www.youtube.com/watch?v=IgJ6SpHZir8

CNN: Electromagnetic Mind Control Weapons (2 of 2)
https://www.youtube.com/watch?v=boZlofptQiw

Dr. Patrick Flanagan on Electromagnetic Frequency Mind Control Weapons
https://www.youtube.com/watch?v=xESAbEllSrQ

Mind Control Experiment Evidence by Cheryl Welsh:
http://mindjustice.org/2003_survey.htm

**No matter how difficult life is today - no matter how much is torn away
The sun will rise on all of humanity as we embrace our hearts and set Love free.**

**Give us STRENGTH, God...to find our way through bullets
hidden in microwaves, and COURAGE, God...to make a STAND**

that saves our lives and FREEs our land.

**PLEASE BECOME AWARE, HELP SPREAD THE WORD
AND FOLLOW YOUR HEARTS ABOVE YOUR MINDS.**

<u>**Heart Over Mind for Human Kind**</u>

Its Worth Fighting For

Humanity's Right to Freely Live and Love and Laugh and Cry.

Its Time to Fight Like Heaven!

I am still deeply concerned that the covert "rescues" are actually enslavement and that technological modes of "protection" could be a sly enslavement performed by those who may also be placing filters in detection technologies, in order to prevent detection of the low frequencies that are used for mind control. Help spread the word on this.

This is a world being secretly torn apart through breaches of mind and blocks of heart but Light will reign as pure as snow to save us from the final blow. Please help it to.

Microwave Weapons

I'm no technological expert. But my experiences with being targeted have helped me to realize a few things about the technologies that are being used on me as well as others whom I know. I hope this article inspires you to do your own research, because this is a critical (literally holocaustal) situation, which needs the undivided attention of governments and citizens around the globe.

Some microwave and laser weapons are classified as "non lethal weapons." There has been much controversy over this being a false classification, due to some of their applications being extremely lethal.

Some laser technologies are not even classified as weapons, although they can be used as lethal weapons. These have been developed for medical purposes with the intent of effecting brain function or performing laser surgery...etc.

Around the early 1990s I read a medical report, which described a new technology that would enable laser surgery to be done, through satellites, on a patient in their own home. Oddly I have not been able to find anything like this article on the web, recently.

There are many scientific reports which show that the medical fields of psychiatry, laser surgery and brain diseases have used radio wave technologies in their treatments and aims to find cures since before the 1970s. Most of the technologies used were obviously invented to help people, but criminal use of them can do as much, of not more, harm than any lethal weapon.

Criminal use of these sorts of technologies were reported in the 1970s when the MKULTRA program was publicly exposed. But I do not think that MKULTRA was the only program or source of these sorts of crimes, which now appear to be for the purpose of harming, torturing or gaining control over human beings through interfering with or controlling brain function...etc., and no longer just for "experimentation."

Some reports state that the types of microwave and laser weapons, which are used on Targeted Individuals, are locally operated from nearby buildings or vehicles. Some state that it is being done through cell phones, computers and other technologies being flooded with unusual concentrations of radio waves microwaves. Some state that the beams of radio waves (microwaves) are being directed through satellites and/or through the types of technologies that bounced radio waves off of the ionosphere in order to be redirected to a distant target. Some state that people are being targeted through a network that utilizes things like cell towers. And some cover the crimes and try to make us look crazy for standing up against them.

Common sense tells me that any technologies, which can emit and direct radio waves can be criminally used. If satellites can direct radio waves to a person's home or computer for internet access, they can also direct them, in more intense frequencies, at a person. Many of my experiences have proven to me that this is indeed happening (VERY remotely) and that it is being done in conjunction with satellite surveillance systems, no matter what other technologies are being used.

Research can be difficult, because privately owned laser technologies, and their criminal applications, are not going to be publicly advertised. And neither are those that were designed as weapons for secret military use. But I hope you let your Heart give my testimony, and that of other targeted individuals, the benefit of your doubt, because we are being hurt and are in need of people believing us and protecting us from further harm. This could benefit you as well, because criminal use of radio wave technologies can only grow and get worse until it is openly detected and stopped.

Microwave expert, General Barrie Trower describes Microwave Weapons secretly being developed and used on humanity since the 1950s. He describes them as military based technologies, which function in the same way as the Russian SURA and the American HAARP, which bounce radio waves off of a layer of the earth's atmosphere, in order to redirect them to their target. It is reported that there are at least six of these types of technologies scattered around the world. And there are probably many other types as well.

On the subject of their use on humanity General Trower says that a person or country can be beamed/targeted from as far away as the opposite side of the world. . .and "By changing the pulse frequency... of the microwaves going into the brain and interfering with the brain. . . you could induce psychiatric illnesses to the point where a psychiatrist could not tell if it is a genuine psychiatric illness or an induced psychiatric illnesses." and "Its very easy these days to radiate them and have them wind up in a jail or in a psychiatric hospital." and "You can beam people. . .to give them cancer, breast cancer, neurological illnesses. You can chose what you want them to get. . .the speed that you want them to become ill."

Microwave Expert General Barrie Trower
https://www.youtube.com/watch?v=kvn-8ITy0oc

In 1976 Time Magazine wrote, "Last month the U.S. confirmed that for some 15 years the Soviet Union has been beaming microwaves at the hulking nine-story U.S. Embassy..." Many employees of that embassy were reported to have become terminally ill with things like leukemia, which was attributed to the microwaves.

There are many different ways that radio waves can be used to effect the functionality of the human brain, organs, muscles, tendons and nervous system. Their use on heavily Targeted Individuals includes physical harassment or torture, psychological harassment or torture and mind control.

Many victims seem unaware, especially of the mind control - brainwashing part, which is less obvious. But many report experiencing "electronic harassment" and "V2K" - the military term for "Voice to Skull" transmissions of sound. And sadly, most victims are not believed by loved ones and professionals who are not aware of the existence of these weapons or their use to experiment on, control, harass, torture or inconspicuously murder Targeted Individuals.

Symptoms can vary, depending on how a person is being targeted, and are difficult to list, because beams of radio waves shot into various parts of the brain or body can create a large variety of false symptoms or actual illnesses and injuries.

A painful result of inconspicuous microwave and laser weapon attacks surrounds the doubt of our testimonies. Some victims have been being falsely diagnosed as "mentally ill" - with things like paranoid schizophrenia, hypochondria...etc. The psychological and physical tortures, which are happening to us need to be exposed and stopped - victims need to be protected from further harm and given the opportunity to understand what has happened to them so that they can recover.

Many may question how we can know if our pain is natural or technologically induced. Heavily targeted victims - those of us who also experience covert harassment (gang stalking) are sometimes taunted by our abusers, who actually let us know what they are

47

doing, in the form of covert threats before the attacks happen. We KNOW that we are being attacked, with microwave and laser weapons, and we need you to at least give us the benefit of your understandable doubt unless it is proven to not be true, through honest medical testing for cell structure damage...etc., and radio wave detection technologies. . .and NOT through the harmful pharmaceuticals, which have been pushed upon heavily Targeted Individuals, in order to determine if the symptoms are from "mental illness" or not. The medicating can not result in an accurate determination, because those who target us can just back off as soon as the drugs and "mental illness" labels have been applied. The psychiatric drugs actually aid the mind control part of the targeting. The whole process of falsely labeling and medicating victims appears to be part of the targeting process.

There are technologies, which can detect or block microwave weapon intrusions. But possession of the blockers is illegal in most countries. And I have read a professional report, which stated concern about new detection technologies possibly having a filter built into them, which prevents detection of the low frequencies, which are used for the mind control part of the targeting.

Criminal use of microwave and laser weapons is probably the most dangerous thing humanity has ever had to face and stop. Scientists have expressed a concern that the type of microwave weapons, which bounce radio waves off of the ionosphere, in order to redirect them to their target, are the true cause of "global warming." The damage they can inflict on human beings ranges from brainwashings and psychological torment to physical injury, illness or death. When linked with covert psychological harassment (gang stalking) it becomes even more lethal to the Targeted Individual.

Among the names and applications of the technologies, which are used to effect the human brain, are; Psychological Warfare, Psychotronic Weapons, V2K, laser weapons, Microwave Weapons, Behavior Modification Technologies, Synthetic Telepathy, Acoustic psycho-correction, Satellite Terrorism, Remote Neural Monitoring, Voice of god Weapon, Radio Frequency Weapons, Bio-communications technologies, Radio Wave Mind Control, Electroenergetics, Electronic Harassment, Directed Energy Weapons, Brain Warfare, Geophysical weapons, Mind Machine, Psychological Language Machine, Acoustic Heterodyne Weapon, Embryonic Holography, Optogenetics, Electromagnetic Stalking, Bio-communications Technologies, Radio Wave Mind Control, Microwave Mind Control, Bio-Electromagnetic Technologies, Bio-energetics, Electromagnetic (EMR) mind control, Electroenergetics, Geophysical Weapons, Psychoneurological Weapons, M.I.N.D. - Magnetic Integrated Neuron Duplicator, Psychic Warfare, Acoustic Heterodyne...etc.

I feel that the mind control and pharmaceutical/psychiatry parts of the targeting are the most dangerous, because it interferes with our natural process of personal and spiritual growth. When we are not allowed to retain our own natural process of thinking and feeling we can not grow and evolve into all that we were meant to be. For years now, my soul has been crying STOP - PLEASE STOP! But I am still being targeted. And many people, whom I know and love, are still being targeted. . .some are being as physically and psychologically tortured as me. Our pain and suffering is immense. We are in desperate need of protection from further harm.

"The ultimate tragedy is not the oppression and cruelty by the bad people, but the silence over that by the good" ~ Martin Luther King Jr.

PLEASE DO ALL THAT YOU CAN TO HELP STOP CRIMINAL USE OF MICROWAVE AND LASER WEAPONS

Chapter Four

Covert Harassment

Its Worth Fighting For -

Humanity's Right to Freely Live and Love and Laugh and Cry.

Its Time to Fight Like Heaven!

I am still deeply concerned that the covert "rescues" are actually enslavement and that technological modes of "protection" could be a sly enslavement performed by those who may also be placing filters in detection technologies, in order to prevent detection of the low frequencies that are used for mind control. Help spread the word on this.

Covert Harassment (Gang Stalking)

Exposing the Secret Criminal Part of Our Own Societies
(Psychological Warfare and the Enslavement of Humanity)

I'm no expert on this, but I am a long term victim of covert harassment. . .and have been doing my best to gain an understanding of it, with the hope of helping to expose it and bring it to an end. I may not have it all down perfectly, but this is the best I can do while still being targeted. Please listen to your heart and let this statement lead you into doing some of your own research and then doing all that you can to help expose it and stop it from growing and continuing.

The most common name for this crime is "gang stalking," which implies that the criminals are like a rough street gang. This is misleading, because it isn't that sort of thing at all. Most members of the covert harassment program are respected community members like waitresses, nurses, government employees, teachers, lawyers, construction workers, doctors...etc. Most members of the program appear to be mind control victims. Many play such a small role, in the targeting process, that they are probably not aware of being used to help destroy the lives of their fellow human beings. Some appear to be victims of technological targeting who were "rescued" into enslavement. . .sometimes in ways that leave loved ones to think they are either dead or missing. On these darker levels of it some members are obvious criminals or dark occult members who are used to execute crimes against their victims. Every level of it appears to be lead/controlled by some sort of dark occult. And the key to its success appears to be pharmaceutical and technological mind control.

There is a desperate need for public awareness of the covert targeting and recruiting/enslavement of good decent people around the globe.

Judging by the testimonies of victims, there appears to be different types of harassment programs that are implemented for different reasons. Some people seem to suddenly become victims due to a member's vengeance. Some appear to be targeted for the purpose of recruitment. And some are long term victims of what appears to be sadistic psychological experimentation and torture, which also includes technological targeting.

Most of the general "gang stalking" methods seem to be basically the same for both long and short term victims. We are stalked and harassed. . .literally every aspect of our lives infiltrated by stalkers (puppets) who are lead by those who obviously hold us under surveillance and sometimes also use microwave and laser weapons on us.

A Hint of Covert Harassment Exposed in the News;
http://vimeo.com/63045047

As far as I know, the state of Michigan is the only state in the USA, which has recently acknowledged the devastating effects of covert harassment and has passed new laws to help prevent its growth and success.

Being a victim of covert harassment is truly a holocaustal hell. I have been being heavily stalked and harassed, while being technologically tortured. I can attest to the fact that being a victim of this is a hell, which no human being should ever have to experience. The covert stalking and harassment process is so cruel and barbaric that it becomes unbelievable to those whom we try to get help from. Our loved ones are literally brainwashed against us. . .and sometimes even recruited into the program. Perhaps the most painful part is the fact that there has been no safe place for us to turn to for protection and help. We are being hurt in ways that are indescribable. Our lives are being destroyed. We are in desperate need of protection from further harm. There is a desperate need for public awareness of ALL levels of the targeting and recruiting process.

There is a desperate need for humanity to understand what is happening so that victims of enslavement can be set free and Targeted Individuals can receive the acknowledgement, support and protection that has been too painfully missing.

When I started to realize that I was being targeted, (in 2005) it vamped up and turned into a push to convince me, and anyone whom I was close to, that I was just "mentally ill". Until I learned about the technologies they use on me I struggled to retain some level of trust in my own sanity. There are many times when I thought I may have been going crazy. The confusion and agony I've experience is truly indescribable. Those years of being surrounded by such cruelty and chaos, while not fully understanding what was happening to me, were (in some ways) the most difficult for me. And I'm sure they were no picnic for my loved ones who have also been targeted in different ways. . .heavily through brainwashings against me. I have no doubt that their pain ran as deep as mine, as we were torn from each other, although they have not been allowed to believe that the targeting is really happening.

In the shadows of the secrecy around the covert program, and it's inflictions of technological mind control, family members are being deprived of each other's Love and support. Pain and confusion fills those gaps where understanding Love needs to be growing.

Are you a Puppet? Are you in the covert program?

If so, you are most likely a mind control victim who is being used by criminals in organized crime and psychological warfare against your fellow human beings. Your part in it, no matter how small it may seem to you, is helping to destroy people's lives. Some of you even seem to think you are fighting for Freedom while you are being used to help to destroy it. Please stop the covert war.

All members of the covert program, BOTH common citizens and government officials; I beg you, for your own wellbeing, and that of your loved ones, as well as our country and the rest of humanity, please refrain from participating in the covert program. Please OPENLY UNITE into a PEACEFUL STAND to help restore the precious Freedoms that have already been lost by too many.

Please Follow Your Own Heart Instead of the Program

51

A Few Examples of Covert Harassment:

This list is just a few of the things that the stalkers (puppets) to do to us. Targeted Individuals often seem to have different experiences, because the targeting seems to be geared toward things that will trigger emotional pain, anger, fear...etc.

Stalking us Everywhere We Go: Stalkers seem to be connected to a large occult/organization, that has chapters in most communities and countries, and pass us off to each other so that the harassment never ends. . .no matter where we go. I have found it shocking how many people are being recruited into this covert program.

Invade our Privacy: Stalkers tap into our computers, phones, emails, homes, relationships, jobs...etc., and relentlessly interfere, sabotage, harass and degrade...etc.

Crowding us in Public: Puppets aim to cut us off or stand too close or bump into us or make us wait in long lines. (This part is also called mobbing.) Stalkers also follow us in vehicles. . .often pulling out ahead of us instead of behind us. They try to cut us off, blare horns at us, pull up beside us with blaring music, try to run us off the road, suddenly slow down in front of us. They sometimes use false license plates on their vehicles when obviously zooming in to cause harm or severe irritation. Otherwise they try to blend in with other traffic and just be inconspicuously rude. I had an experience with one who must have had some sort of switch to shut off his license plate light, because as I tried to read it the plate light went out and he sped off. They often turn vehicle lights off or on or high or low when approaching us. They have often done things like banging on public bathroom doors as soon as I sit on the toilet. I've experience repeated rounds of street lights being turned off and on as I enter a parking lot or street...etc. This appears to be to let us know they are there. I guess its supposed to be intimidating or scary.

Home, Business and Vehicle Intrusions: They access our homes and vehicles, in order to move our belongings around...etc. I had often returned home to find things like my picnic table moved from one side of my house to the other in the early 1990s. They play games like hiding something one day and then putting it back after we notice it missing. I have experienced them repeatedly moving the side and rear view mirror and the seat in my vehicles...etc.

"Street Theater" - Noise Campaigns: Puppets create frequent loud noises around us, which can include things like blaring music, blaring horns, banging, slamming doors, blaring sirens, yelling, revving engines, screeching tires. . .and street theater - the process of acting out an argument between two puppets, which often includes the repeating of parts of our own conversations...etc. A lot of it is crude and contains a lot of swearing and negative messaging.

Financial Ruin: The puppets instigate financial ruin. This can include brainwashing techniques to steer us in bad directions, identity theft, false rumor campaigns and the sabotaging of jobs, which can include the inconspicuous destruction of our equipment, interference with phone messages, emails and text messages...etc. It can also include microwaving us until we are too sick to work. I have experienced extreme levels of this until they had me destitute and homeless. This list could go on and on and on.

Sabotaging of Relationships: Puppets are used to help destroy our relationships with our children and family members as well as our friends and coworkers...etc.,. . .through false rumors, fabricated letters, fabricated emails and what appears to have been fabricated recorded phone messages and technological mind control intrusions...etc.

Public Slander: Puppets are used to pass false rumors in communities we move to, on the internet or in the media, in order to publicly discredit us. I have heard of them doing this to me, but have not directly seen it. A news team puppet had once laughed and told me of a news cast reporter who had said, "I was afraid of what they would think of me," which was a cutting joke about something I said in front of TV cameras, at my neighbors funeral, directly after my neighborhood had been wiped out in a flood and I'd been drugged and raped and brainwashed into thinking that the deaths of my neighbors were my fault. Someone also told me that I was on the weather station's "Storm Story" about the Alstead flood, and that they portrayed me as a "Joan of Ark." I guess they ridiculed me for a while, but I didn't subject myself to it. I never saw any of it. But it hurts that people can be so cruel. I really did blame myself for the deaths of my neighbors. The pain all of this inflicted was devastating to me. And I didn't start understanding it until I realized that I was being targeted. For a while I really thought I was going crazy.

Sabotaging Vehicles: I have experienced unusual amounts of sudden flat tires, sudden brake failures, electrical problems, batteries being drained...etc. Air is often let out of my tires. My vehicles have repeatedly developed oil leaks, due to the oil filters being unscrewed or bolts suddenly missing from the engine. They have often sprayed something on my windshield, which makes it nearly impossible to see out of when it is wet. This is a substance that is difficult to wash off. It appears that they also spray something on vehicles that speeds up rusting and leaves a yellowish film on the vehicle. It appears that they have put salt into my car doors to make them rust out. When we have older vehicles they can have a field day and just write it off as being due to the vehicle being old. The goal seems to be to either instigate an accident or destroy our vehicle in ways that appear accidental or like normal wear and tear.

Threats: Threats are delivered in cryptic ways that repeatedly mention or display death. "Happy Birthday" appears to be a covert threat. Stalkers will come close to me and yell out a message, while pretending to be talking to someone else. . .in such a way that I know the message is for me. (One obvious happened when a man pulled in next to me, stood next to my vehicle, pretending to talk to his partner and yelling, "You keep testing! You know what is going to happen if you don't STAND DOWN!" This was while I was testing to see if microwave weapon attacks were coming from vehicles (like they want us to think) or from a more remote source, like satellites. My conclusion was a FIRM knowing that I was being attacked from the sky - most likely through satellites and this angered them.

Electronic Harassment: The electronic harassment hits many levels and different arenas and could fill a book. Apparently it can happen VERY remotely or be done locally with various types of microwave or laser weapons set up near a victims home. Most of the attacks I experience seem more remote and probably are due to my habit of doing a lot of moving around.

What appears to be the more localized attacks, performed by puppets, can range from being uncomfortable to being extremely painful and disorienting. Sometimes its heavy blasts of microwave energy and sometimes its like being shot with some sort of laser weapon, which causes an instant rush of heat and pain...sometimes only pain or a strange, intense, fast vibration type of feeling. I've experienced sudden unexplainable inflictions of what appears to be burn marks on my body.

Biological and Chemical Warfare: The most criminal types of puppets use chemicals that burn our eyes, nose and lungs or cause itching and rashes and physical illness. I have experienced chemicals put in my car, shoes, clothes, on toilet paper...etc. I've also experienced suddenly breathing in something that burns my lungs and makes me choke. In December 2010 I experienced sudden, intense pain in my lungs and choking up globs of mucus, which had little black dots in it. . .at a time when I was not ill.

Stealing or killing of victim's pets: During times of heavy rounds of targeting I've had missing pets and a dog that died from a medically unexplainable infliction, which was probably remotely done with a laser weapon. I believe that my pets have also been targeted when they were allowed to remain with me.

Inconspicuous Murders of Loved Ones: Long term victims, like myself, can be surrounded by an unusual rounds of deaths, which FEEL unnatural although most of them appear to be of natural or accidental causes. I have experienced this. Some deaths appear to be distractions from other realizations of these crimes. Those who target us appear to have no regard for life.

Psychological Harassment and Torture: Puppets say and do countless numbers of things to confuse or scare us. Some just seem like rounds of foolish games, like the time when they said, "We are all wearing sunglasses" and then surround me with puppets who are wearing dark sunglasses and are obviously staring at me. I guess its supposed to intimidate me.

At other times they do rounds of having the puppets slam their vehicle doors around me. I've parked in places where I am suddenly surrounded by up to a dozen people who are obviously slamming their doors shut, some of them doing to two or three times on the same vehicle. This form of harassment can include painful levels of mental abuse and almost any array of cruel or psychologically confusing scenarios.

They also play catty games. . .like telling us what kind of harm they are going to inflict, through the puppets, just before they do it. This list could go on forever. One example of this happened in 2002 when a puppet came to me and said that his mother had just called him and said that his little brother could die. . .and two days later my little brother died in a mysterious "accident." Unfortunately I did not know I was being targeted or that this man was a puppet, at that time. But I felt that my little brother's death was no "accident" and still do.

Manipulations: Puppets manipulate situations to make us think innocent people are the ones who are following or harassing us. This makes us look crazy if we accuse them. The set ups for us to place blame on innocent people is a STRONG pattern. And I believe that there are massive set ups to make us blame the USA Government in order to pull suspicion away from those who target both citizens and USA Government officials.

Sleep Deprivation: Puppets launch noise campaigns to try to keep us from getting the rest we need. This can also be done with technologies. I have experienced some periods of this, but not nearly as much as other Targeted Individuals report. The last time I experienced this, it appeared to be done with technological intrusions into my brain, because there were no sounds keeping me awake. This was at a time when they had forced me to go see A county social worker, in exchange for a room rental for a couple months. . .which all ended up appearing to be an attempt to frame for being mentally ill.

Parasite Infestations: Puppets can infect our food/bodies, vehicles or homes with parasites. I experienced this, in my homes, in 1995 and 2006. I have occasionally felt or sense a stalker who comes to either sit or stand behind me, putting something on my head or down the back of my neck. . .and then I have head lice. I have caught this in action. In 2012 and 2013 I was hit with some sort of intestinal parasites. (A puppet even called me to let me know, WHILE I WAS drinking the frappe they did it to!) Heavy doses of raw garlic cured it.

Destruction of Homes and Businesses: Puppets inconspicuously destroy our homes and businesses. My first home was taken, in a VERY unusual way, by the New Hampshire Department of Transportation, under their "Rights of Eminent Domain." My second home was destroyed in a suspicious fire. Another neighborhood was wiped in a suspicious flood. And my businesses have been so sabotaged that I've had to aim for other jobs, which were also repeatedly sabotaged or used as avenues to inflict more pain.

Dead Animals Left in Roads: I have experienced many rounds of threats during times when an unusual amount of dead animals were on the roads I frequented. On a few occasions they were left on my doorstep. A puppet had brought this to my attention in 2005. "A satanic occult kills the animals and leaves them in the road," he said. This was around the time when they started forcing me to realize that I was being targeted.

55

Framings for Uncommitted Crimes: Stalkers try to set us up to be framed for crimes. This appears to be to discredit us and seems to lean heavily toward the most inexcusable crime - child molestation. . .as well as theft. I have experienced periods of them literally following me around with children and obviously trying set me up to be framed for sexual abuse. I never have and never would harm a child, and this is what makes it feel so horrible. "What did you do to him?" one of them had yelled out as they were sending boys to where I was sitting or standing...etc. On three recent occasions I experienced obvious puppet attempts to send their child over to climb onto my lap in a restaurant. The poor kids!

They even try to set us under the guise of covert "help." They have disabled my car and then tried to make me steal vehicles, while pretending it was there for me and was "help." I cannot count the amount of times that they have pushed me into destitution and then had store attendants head into the back room or bathroom, as they saw me coming in, and then wait to see if I'd steal something. They have left a purse in a bathroom to try to frame me...etc.

There have been repeated attempts to frame me for being mentally ill through lasering my brain at strategic times or setting up targeting scenarios and then having a puppet encourage me to go seek law enforcement help. (On these occasions, when I sought help they tried to tell me I was "mentally ill.")

There appears to have been repeated attempts to frame me for something to do with the internet. During these times, I got weird feelings each time a box would pop up on my computer, which warned of an IP address conflict - of someone else on that public internet with the same IP address.

There even appears to have been attempts to frame me through pretending to be covertly trying to help me win the lottery. I could fill a book with the obvious attempts to frame me for theft, pedophilia, plagiarizing, murder, mental illness...etc. They have not succeeded because these things are too foreign to my nature. I am not a criminal and I have done my best to listen to my intuition, instead of falling into their traps, but its sometimes like a full time job and VERY difficult when they have microwaves aimed at my head and are effecting my memory...etc.

Inconspicuous Murder: Stalkers may try to kill us in ways that look like a suicide, accident or natural death. It appears that heavily Targeted Individuals are also being slowly killed through toxifying food and directing harmful doses of microwave energy at us, shooting at our hearts with laser weapons, trying to instigate "accidents"...etc. I have even experienced them most likely putting a bullet in a gun, which I picked up and THOUGHT was empty. I was lucky to have been the one to find it, and not my daughters or their friends.

Abduction and Recruiting: It appears that some victims are abducted after being completely isolated from loved ones. There has been repeated attempts to do this to me. A WORD OF CAUTION TO TARGETED INDIVIDUALS; Stalkers will zoom in and pretend to be covertly "helping" us out of the situation. The deceptions around this are HUGE! Deaths are even staged in order to recruit people! My instincts have shown me that we end up in the program and completely controlled if we go with them. I know several people whom this has happened to and the covert program appears to be literal human enslavement. . .and this has helped me to feel compassion for most of the puppets who were probably decent people before being targeted, rescued, enslaved and controlled by the criminals who were targeting them to begin with.

I've been approached by a puppet who tried to convince me that it was a good thing. I've been approached by a puppet that deceitfully tried to recruit me to seek revenge on those who target us. I've been offered money to go with them. And I've been offered freedom and a whole new life. All of it has felt like criminal coercions although it often seems to use decent people in the foreground of the program.

Drugging and Raping: This can repeatedly happen to both the victim and his/her children. It seems to be a sadistic domination tactic. (I had two beautiful daughters and believe that they have both experienced this. One had bruises to prove it. And I hate to admit it, but I have been drugged and raped at least 4 times.

And the gory list goes on. . .

PLEASE LET YOUR HEART REALIZE WHAT IS HAPPENING, LET YOUR COURAGE STAND UP TO HELP END THESE CRIMES, AND. . .

I am still deeply concerned that the covert "rescues" are actually enslavement and that technological modes of "protection" could be a sly enslavement performed by those who may also be placing filters in detection technologies, in order to prevent detection of the low frequencies that are used for mind control. Help spread the word on this.

"The ultimate tragedy is not the oppression and cruelty by the bad people, but the silence over that by the good" ~ Martin Luther King Jr.

The Recruiting Process

I've been fighting to survive while being technologically targeted and while a covert war has been raging around me. I am deeply concerned that innocent people are being harmed by all sides. There is a desperate need for realization of the recruiting process - enslavement into the program that has been taking over the USA since at least the 1970s. I have experienced aims to recruit me in several different ways. I have no doubt that this has been happening to other people. And I feel that it is in desperate need of being exposed and stopped.

I feel that if the whole public were aware of the recruiting processes, and that joining means being enslaved, most people would not participate. . .and the program would stop growing.

I am still deeply concerned that the covert "rescues" are actually enslavement and that technological modes of "protection" could be a sly enslavement performed by those who may also be placing filters in detection technologies, in order to prevent detection of the low frequencies that are used for mind control. Help spread the word on this.

I am deeply concerned that, in this covert war, victims of deceitful and forced recruiting may be seen as the enemy by those who may also be victims of brainwashings. The chaos and suffering that is being caused, by all sides, is holocaustal and bad for EVERYONE.
Please become aware of the following recruiting techniques that are being used on us. And PLEASE help spread the word, especially to government officials.

<u>*It appears that something has been being done to people, in order to protect them from technological mind control - but that "protection" is really a sly enslavement. Both government officials and citizens appear to be victims of this</u>.

* It appears that some citizens are being brainwashed, apparently with the help of mind control technologies. . .and then coerced into the covert program under the guise of it being a good thing. Sometimes people are targeted - pushed into financial ruin and then enticed to join for money. I have experienced these sorts of attempts to recruit me. Victims of this ARE NOT the enemy. Those who deceive them are the enemy.

* It appears that heavily targeted victims are physically and psychologically tortured and then "rescued" after forced isolation from loved ones. This "rescue" is really an abduction. I believe that some "deaths" are even staged, during abductions. I have experienced repeated attempts to rescue me directly after experiencing severe rounds of electronic and psychological torture. And I strongly felt that the "rescue" was from perpetration and not a good source, although they seem to sometimes use good people in the foreground of this operation. Apparently, once the rescue takes place we become a slave to the rescuers. Victims of this ARE NOT the enemy. Those who torture, abduct, brainwash and enslave people are the enemy.

* It appears that when we can not be deceived, coerced or tortured into the program, we are more heavily targeted and become "Targeted Individuals" who are lured into groups where we are brainwashed into blaming only the USA government and are offered opportunities to seek vengeance on "those who target us", which I believe is a sly way of recruiting us into a program that is really lead by those who target us and would use us to target other innocent people. Victims of these recruitings ARE NOT the enemy. If they were aware of the deceptions they would resist it.

* It appears that some of us are drugged and staged for photographs or framed for a crime. . .and then pushed to "escape" with them. I have experienced them telling me about bad pictures on the web around times when they were trying to "rescue" me. I never checked for pictures, but know that if there are any, they were totally fabricated. They have also told me that I have already been framed just before pushes to make me escape with them. The goal, with both of these things, seems to be to make me feel that my life is ruined and that it would be best for me to start a new one with them.

* It appears that covert attacks on isolated Targeted Individuals are being followed by a staged "rescue" that is really a recruiting into enslavement. This has often happened to me. The last time was on the night after March 11, 2015 when I was woken in the middle of the night with what seemed like a microwave attack, which caused sudden fluid accumulations in my lungs and lasering of my brain. Then this morning puppets (one with oxygen tubes in her nose) talked of hospitalization and a push for another covert "rescue". . .and I strongly felt perpetration in the background of it. These sorts of things - attacks followed by pushes for a covert "rescue" have been happening a lot since June of 2013. It was probably happening before that, but I just didn't notice, because I had gotten the message, "Your daughters will be OK if you leave" in 2006.

* On the darker side of recruitment; it appears that some join "the occult" and participate in the gang stalking part as if it is a cool game. If they were aware of what they are following, and that they are being used to destroy the lives of fellow human beings, most of them would not be recruited.

* I hear that some get recruited through prisons where they are trained and then do gang stalking as a job when they are released. This sounds like a dangerous

type of reform that needs some reforming!

The bottom line really is that. . .until huge levels of public awareness are gained, unaware victims are being hurt, recruited, or "protected" by those who are destroying humanities Freedom. In the lower stages of this war, it appears that harmless people are being targeted by enslaved victims who seem to think they are fighting to regain Freedom from the infiltration while destroying Freedom. . .and the fact that this does not make sense should show you how senseless the covert war is.

Meanwhile, the REAL criminals freely continue with their rapidly growing operation, which thrives on the secrecy that ALL sides sustain for them. How much sense does that make?

I feel scared for those, whom I know to be decent people, who have been literally tortured and brainwashed into something that they do not even realize is criminal. . .and for victims who are frantically fighting for Freedom in ways that are helping to destroy it, as well as victims, like myself, who are suffering indescribably through attempts to literally torture us into submission. Its all just too horribly sad.

THIS IS A PLEA FOR OUR GOVERNMENT AND MEDIA TO EXPOSE THESE CRIMES AND STOP CRIMINAL USE OF HUMAN BEINGS AND RADIO WAVE TECHNOLOGIES - TO FREE THE SLAVES, AND ULTIMATELY ALL OF HUMANITY, FROM THIS TECHNOLOGICAL HOLOCAUST.

PLEASE EXPOSE AND STOP THE COVERT WAR. PLEASE!

If you are in a covert program, STOP - Just PLEASE STOP and help save yourself, your fellow citizens, our country and ultimately all of humanity from further destruction. Please follow your Heart instead of them.

Heart Over Mind for Humankind
www.heartbud.com

"The ultimate tragedy is not the oppression and cruelty by the bad people, but the silence over that by the good" ~ Martin Luther King Jr.

Chapter Five

Targeted Individuals

Its Worth Fighting For -

Humanity's Right to Freely Live and Love and Laugh and Cry.

Its Time to Fight Like Heaven!

I am still deeply concerned that the covert "rescues" are actually enslavement and that technological modes of "protection" could be a sly enslavement performed by those who may also be placing filters in detection technologies, in order to prevent detection of the low frequencies that are used for mind control. Help spread the word on this.

Plight of Targeted Individuals

We are unheard victims lost beneath the lies.
We are the fading ones put on a list to die.
We are rising wounded begging for your aide
Becoming specks of dust in an evil charade.

There are different types of Targeted Individuals. I can only speak for those of us who are long term victims of both covert harassment and criminal use of microwave weapons, because this is what I've experienced. What I've written on the other pages of this site describe most of what we experience, especially the pages on covert harassment, Microwave Weapons, mind control, Surveillance and My Testimony

We are held under constant satellite surveillance while being stalked, harassed and inflicted with various types of microwave and laser weapon attacks. The targeting starts out so inconspicuously that we do not realize we are being targeted until those who target us have succeeded with isolating us and instigating financial ruin. Prior to this point we can think we are just having a lot of bad luck. Once this point is reached we are targeted in ways that make it look like we are just "depressed," or have "Paranoid Schizophrenia" when we try to seek support from loved ones or help from law enforcement, especially when they are not familiar with the types of targeting we are experiencing. Most people are not yet aware of the technological part of targeting or the covert manipulations that surround us. Many doctors are happy to try to label and medicate us, instead of performing tests to prove the targeting. I have read reports about doctors who have aimed to prove the targeting and then got targeted or threatened for doing so. Efforts to prevent us from getting the types of help we need are strong and sometimes lethal.

Our relationships, homes and businesses are destroyed or sabotaged. Even our own family members and friends are brainwashed against us and/or are convinced that we are just "mentally ill." Set ups to discredit our testimonies start in the early stages of the targeting and can include false rumor campaigns (slander), fabricated emails, letters or phone messages and literal technologically induced brainwashing.

Consequently we are often forced to either not seek help or struggle to accomplish the seemingly impossible task of proving our sanity - proving that these technologies exist, and are being criminally used. . .and then proving that it is actually happening to us, WHILE WE ARE BEING TORTURED WITH MICROWAVE WEAPONS! This has had no success, that I know of, due to pub-

62

lic lack of awareness, our diminished functionality while our brains are being microwaved and interference by those who target us.

I've experienced good objective reactions from some law enforcement personnel, which I have deeply appreciated, although they were not able to help protect me. And I've also experienced some who have launched into intense efforts to try to convince me that I am "mentally ill."

Sadly, some Targeted Individuals have been shoved into institutions. Can you imagine "being sane in an insane place"? I know of a woman whom this happened to. And it tears at my heart, because it almost happened to me as well. I want to find her and save her, but my hands are tied and this feels horrible. There are times, like when I remember my daughters being brainwashed into trying to have me institutionalized, that my heart has cried. . .

I don't want to be left to evil pretenses of helping hands.
I need to be comforted by those who can care to understand.
I don't want to be declared insane for their hateful gain.
I need you to soothe my wounds instead of inflicting more pain.
I don't want you to watch from a silent distance while I die.
I need you here beside me as I pray to God and cry.

No problem ever got solved by misdiagnosing it!

Much of the targeting seems literally satanic and inflicts painful rounds of psychological and physical torture. There seems to be a heavy aim to instigate feelings of being unloved while being tortured. There are many reports of heavily Targeted Individuals committing suicide and this appears to be one of the aims of those who target us. But these are not REAL suicides - they are psychological and physical murders that are called suicide. They also try to frame us for crimes in efforts to have us incarcerated or discredited/slandered. And then there's the "rescue" - the part where they attack and set us up to be framed and then zoom in pretending to be rescuing us. I believe that the only reason this has not worked with me is because I have publicly exposed it and have not fallen for it. But, sadly, it appears that many have. I strongly feel that the "rescue" is really a disguised abduction, which they can call our choice to be shoved into literal enslavement with the use of "mind control technologies. I believe that this has happened to some of my loved ones. And it is too so sad, because they are no longer who they were.

After we are isolated from loved ones its more than difficult to form new friendships. When we get close to an individual they are either brainwashed against us or targeted into such states of emotional overwhelm that they have no energy left for us. And this appears to be the intention. In my case, there has been many deaths. Inconspicuous murders of the loved one of someone who is starting to believe us or care about us is not at all uncommon. I have been reaching out to people and am recently realizing how severely the targeting fol-

63

lows me. . .and this now has me feeling like I am trapped. I don't want people to be hurt for listening to me. Yet without public awareness the crimes seem to freely grow! Its a horrible "catch 22!"

I understand how difficult it is to believe that this is really happening to people, but it is true. What is happening to us is as inhumane as inhumane can possibly get. We are suffering indescribably - we are suffering in ways that no human being should EVER have to suffer.

Many Targeted Individuals have web sites and blogs on the web. But those whom I know personally are not on the web. I had found one woman on the web - (Racheal Orbin) who seemed to be another long term victim and was going through very similar chains of events as I have. But it appears that we were blocked from connecting and I got only one email from her before she seemed to have vanished.

There are a few people, whom I had been close to, who are also being heavily targeted. One has lost his license to practice medicine and has experienced strange chains of "accidents". . .like a head on collision with a police car and a skiing accident that left him partially crippled. My gut feeling is that these were NOT Truly accidental. The fear and pain I saw in his eyes the last time I talked to him has haunted me as much as that in someone else whom I had been close to before the targeting vamped into lethal levels on all of us. It appears that those who can not be easily controlled or recruited are literally tortured around repeated attempts to force a "rescue" into enslavement. (It appears that those who have been enslaved are used in the "gang stalking" part of the targeting.)

This is a critical situation that is in desperate need of public awareness and government action to free those who have been enslaved and protect the rest of us from further harm.

When I read the following statement it forced my aching heart to release a batch of tears, "I wish to everyone who has lasted long enough to read this, all luck in searching for just normal life. When we will get the right to that, the most important battle of humankind will have been won." by Targeted Individual, Aleksander Zielinski. (I wonder if he is still alive.)

PLEASE LET YOUR HEART FIND THE COURAGE
TO STAND UP AND HELP STOP THESE CRIMES.

My Blog; www.sharonpoet-ti.blogspot.com

"On the Need for New Criteria of Diagnosis of Psychosis in the Light of Mind Invasive Technology" by British Psychoanalyst Carol Smith
http://www.globalresearch.ca/index.php?context=va&aid=7123

P.S. Please do not look for faults in Targeted Individuals instead of helping us in

the ways that are needed. We are surrounded by manipulations that are intentionally set up to make us look bad. The people who target us, and those whom
they control, will use those set ups to discredit us and make people look down
on us.

I have undergone round after round of attempts to frame me or have me
labeled as "mentally III." The "mentally III" labeling stuff started in 2003 - directly
after I reported my belief that my little brothers death was not an accident. Other
members of my own family have been used in this process of trying to label me,
although I have been separate from them through most of my adult life. The aim
has also continued through puppets - people who are controlled by those who
target me. It has often felt like a full time job to just avoid these set ups.

And on the smaller scales; There appears to be an aim to try to make me
look bad for smoking cigarettes. I've caught puppets taking pictures of me smoking. What they will not tell you is the extreme levels of distress, torture and
abuse that they used to drive me back to smoking in 2008/2009. . .and the ways
that they have tortured me with microwaves and/or surrounded me with smoking
puppets every time I try to quit. The last time I quit they even had two puppets
offering to buy me a pack of cigarettes. When I quit smoking I usually get emotional, for a while. . .as I release pains I've suppressed with the cigarettes. And
this now possess a serious problem, because there have been too many times
when I've been tortured for crying or having any sort of deep feelings. . .and
times when they have had people suggested that my tears are just cause to
label me as "mentally ill."

I do not know of any words that can describe how horrible it feels to be held
in a public prison, remotely tortured and not allowed to fully take care of myself.
While being heavily technologically and psychologically tortured, cigarettes have
actually helped me to retain my sanity through numbing out some of the emotional pain that I have not been allowed to privately and freely release. (This
may appear to go against what my original work is about, but my past statement
do not apply to extreme situation.)

While under these extreme levels of distress, and forcibly held in destitution,
I have even bought cigarettes with the gift cards I've gotten from churches...etc.
And that leads me to another set up; The targeting has also forced me into, and
holds me in, such a state of poverty that I've had to seek help from different
towns, organizations and churches. I guess this is something we are not supposed to do, but I have done what I have to survive. The first organization I had
sought help from severely limited degrading levels of help and repeatedly suggested that I was not allowed to get help from any other place. This seemed like
part of the torture process. After that I sometimes felt guilty for seeking help in
other places so that I'd not starve. They changed their tune, with me, after I
started writing about it, but I wonder how many other people are treated this
way. And I still feel like I am doing something wrong for seeking help in other
places.

My homes have been destroyed, in various ways, by those who target me.
My work has been repeatedly sabotaged. And the dozens of other jobs I have
aimed for were either used as a way to inflict deeper levels of abuse or were
quickly sabotaged when they are with people whom they do not have control

65

over.

Since 2004, I have repeatedly aimed to restart my work. . .even changing my name and its name and locations, but it had been repeatedly sabotaged. And the amount of financial help I get seems to have been being controlled by those who target me.

My first reports about something suspicious happening were in 2002 or 2003 after my little brother's death, which happened within a year after he started figuring out that our family was being targeted. I have been begging for every level of personal help since 2006. But those who target me have prevented the levels of help I need.

Since 2006 I'd been held in a state of having to spend most of my time trying to figure out how to survive while surrounded by manipulative walls that hold me in a state of destitution. (I have even begged for money at gas stations and in parking lots.) There are those who chose to use my need for help a reason to judge me, instead of helping me. But, thank God, there are also those whose Hearts understand and do what they can to help. I think of them as my human angels although I do not know most of them.

"The ultimate tragedy is not the oppression and cruelty by the bad people, but the silence over that by the good" ~ Martin Luther King Jr.

Symptoms of Microwave Targeting

This includes both symptoms of microwave mind control and that of more lethal forms of microwaving, which I have either witnessed and/or experienced.

Symptoms of Microwave Mind Control; Most victims of microwave mind control appear to experience unnoticeable brainwashing and are not aware of what is happening. The most common noticeable symptoms are mental numbness and fatigue, a faint high pitched ring in ears, unusual neck tension, unusual head aches and unusual thoughts that do not match feelings.

The most dangerous long term, general effect is interference with our natural process of personal and spiritual growth - the loss of psychological freedom.

Long term mind control victims can become disconnected from their hearts and natural instincts and can seem narcissistic. The interference with our processes of feeling, thinking and sensing is LETHAL to our psychological health. We can not grow and evolve and become all that we were meant to be while our brains are being intruded upon and manipulated.

Long term Heavily Targeted Individuals can also have symptoms like unusual tooth decay, unusual dry scalp, ridges on finger nails, chains of unusual medical inflictions, unusual discord between family members, unusual stress, mental confusion, forgetfulness, unusual or repetitive thoughts and mood swings, periods of diminished hearing and eye sight, sudden onset of unexplainable neurological problems and sharp pains shooting into head. . .as well as symptoms that mimic mental illness. . .primarily "depression," "split personality disorder" and "paranoid schizophrenia." Parts of the targeting seem sadistic/satanic with intentions to inflict psychological suffering and chaos. In severe cases, it appears that unaware mind control victims can be completely controlled by criminals who can force them into suicide or things like lethal shootings at schools, libraries or navy yards.

Some people seem less susceptible to being completely controlled - perhaps those of us who are more creative, independent or strong willed. People who are taking mood altering drugs, like anti-depressants, are VERY susceptible to being completely controlled and these drugs are reportedly being heavily pushed onto people who are NOT "mentally ill" and merely need to face their issues or release their emotional pain...etc.

Our own hearts and natural instincts can override SOME of the manipulations, if we are aware and are not too heavily targeted. BUT if we are not

AWARE of what is happening we can easily confuse the intrusions with our own thoughts, feelings and instincts, and this can be devastating. . .to say the least. Through awareness and a strong will, we can avoid manipulations that are alien to our own nature. But when our existing issues or feelings are triggered, or when we are also inflicted with drugs or heavy doses of emotional trauma, it is far more difficult to resist.

The most recent developments in Mind Control Technologies, which are said to have mind reading capabilities, seem too outrageous and dangerous for most of us to even want to believe. But they do indeed exist and PLEASE believe me when I say that this crisis can only get worse until people become aware of what is happening and take HUGE peaceful steps toward bringing it to an end.

Symptoms of more Lethal Forms of Microwave Targeting;

Symptoms of other sorts of microwave and laser weapon targeting seem to depend on how we are being targeted and for what purpose. They may slowly increase, come and go, or continue almost constantly. It appears that microwave and laser weapons can inflict almost any sort of physical illness or injury, through directing microwaves, or finer beams of radio waves, into various parts of a person's brain or body.

Heavily Targeted Individuals experience symptoms that range from periods of; unusual or repetitive thoughts and mood swings; diminished hearing and eye sight; medically unexplainable head and body pains; diminished eye sight and hearing. . .to painful levels of electronic torture and brain damage.
Long term experiment cases can have unusual tooth decay, ridges on finger nails, chains of unusual medical problems, brain damage, unusual discord between family members...etc.

These are some of the most common symptoms I get, or have witnessed in others. These symptoms never happen all at once. I get sudden rounds of different things, which can mysteriously stop as quickly as they started. Many of them are medically unexplainable and suddenly happen at times when I am being threatened or otherwise obviously being tortured by those who target me. These periods of obvious torture most often inflict me with; severe pain in my head (often with a loud ringing in my ears); sudden pain and/or fluttering in my heart (this has sometimes been accompanied by the puppet messages, "I'm having a heart attack" or "You are giving me a heart attack"; sudden unexplainable pain in spine; sudden rounds of shortness of breath; sudden rounds of fluid accumulation in my lungs; sudden rounds of unusual bowel elimination; sudden rounds of diminished or increased hearing and sight...etc.
Some symptoms. . .like mild ring in ears, ridges in finger nails, dehydration caused by microwaves, mental fatigue, short term memory loss...etc., are ongoing - almost constant.

Sudden bloating of my whole body - Hair loss (more than the norm) - Pain in head (sometimes severe and in different areas) - Altered depth perception (during severe attacks) - Sun sensitivity (during severe attacks) - - Emotional numbness - Uncharacteristic Rounds of Anger - Physical Fatigue (sometimes debilitating. I get this a LOT!) - Sudden Rounds of Nausea - Sudden onset of unusual sinus problems (without infection or cold and with dark circles under eyes) - Surveillance Symptoms: Feeling like I am being watched when there is no body there. - Dehydration: (Often severe even when consuming a lot of fluids) - Shortness of breath (sudden and unexplainable - often with the fatigue) - Dizziness (sudden rounds of it during severe attacks) - Sudden sweating (sudden heat rushes in body or head) - Body feeling feverish but temp way below normal (96 to 97) - Deep pain down right side of back of head (often forming a knot at base of skull) - Sudden bouts of unexplainable heat in head or body - Pain behind right eye and eyebrow (gets intense!) - Unusual weight gain or loss (This can be sudden and severe) - Sudden unexplainable open sores on body - Thoughts that feel unnatural - Gums swelling and teeth bleeding and hurting - Burning sensation in mouth (sometimes a metallic taste - Tingling in body (usually in nose, rectum and pubic area) - Sudden stinging sensations (as if being bit by a bee, though there is no bee) - Pain in joints and muscles (sudden rounds of) - Intestinal problems (including elimination problems from one extreme to other for no apparent reason) - Sudden spells of fluttering in my heart (racing and skipped beats for no apparent reason) - Heart beat vibrating through whole body (often with the fatigue and shortness of breath) - Heart Attack Symptoms (for no medical reason and usually accompanied by threats) - Lupus - Elevated SED rate (Severe in 2006.) - Malfunctioning of technology (Since 1970s, watches computers and vehicle electronics malfunction) - Dreams with Unusual Messages (Psychotronic weapons can project dreams into our brains) - Brain Farts: (Intrusions into our brains can cause interrupted thoughts or speech) - Nerve Irritation or Damage: (Periods of unexplainable numbness in lips, face or right arm. Muscle or nerve twitching in eyes or other parts of face...etc.) - Sudden unexplainable inflictions of parasites (Eating raw garlic helps.) - Brain injury (for no apparent reason) - Cataracts at an unusually young age - Unusual cellulite accumulations - Sudden unusual mental blocks with certain things - Joint and back problems - Sudden Unexplainable Pains in any part of body - Sudden unusual obsessions (either for or against someone or something)

Unusual rounds of anger (usually at strategic times) - Unusual onsets of cravings for sugar and unhealthy foods - V2K (technologically projected voices to make victim appear mentally ill.) - Sexual stimulation that does not feel natural - Seizures: Sudden seizures for no medical reason. - Forced Speech: Saying things and not knowing why you said them and sometimes even sudden episodes of gibberish talk - making no sense (rare) - Personality Change: Slow or sudden altered belief systems, taste, desires, preferences, values, morals ...etc. - Sudden Onset of Unexplainable Physical Injuries. - Things like sprained ankles, joint problems and slipped discs can be remotely inflicted. - Tumorous Growths: These may or may not be cancerous. - Various types of organ failure - Autoimmune diseases - Various types of cancer - Diabetes.

Birth Defects: I have read reports that a targeted pregnant woman can have a child with a minor heart defects. One of my children was born with a minor heart defect and four breasts. I was told by a doctor that I have a problem with a valve in my heart. . .but then another doctor told me that I didn't. Its hard to determine since some doctors are part of the criminal operation. I have recently been told that two of my brothers also have minor heart defects. I fully believe that my mother was a Targeted Individual.

Stunted Growth: (Perhaps the worse symptom is one that is not immediately noticeable - the blockage of our Hearts - the blockage of our ability to think and feel in ways that come natural to us. This is a serious hindrance to our natural process of personal and spiritual growth - our process of evolving into healthy human beings.)

Depression or sudden onset of mood swings (which do not match feelings) Symptoms that Mimic Mental Illness: Victims can experience sudden mood swings and feel anxious - like they know something horrible is happening to them, but can not figure out what it is. Victims and/or their loved ones can easily assume it is a "mental illness" due to lack of awareness of all forms of microwave targeting. And, sadly, some doctors are more than happy to prescribe medications that aid the complete success of said technologies. Hearing voices that are sneakily technologically projected are common. (These voices can also be projected to people who are near the target, in order to make them think the target is saying foul things...etc.) This latter one appears to happen to me at strategic times.

Morgellons Disease: A parasitical illness that appears to be shot into people and/or placed in food or drinks of a targeted person.

Misdiagnosis: Falsely labeling Targeted Individuals as "mentally ill" is almost the worst part of this, because it forces 'help' in ways that are NOT needed and completely prevents help in the ways that are DESPERATELY needed.

Thus far, heavily targeted victims, who have become aware of what is happening, have not been able to obtain help or protection. We are being covertly tortured in the worst kind of prison - a prison that is filled with people who appear to be brainwashed into disbelief. Can you imagine?

We are unheard victims lost beneath their lies.
We are the tortured ones put on a list to die.
We are rising wounded begging for your aide
Becoming specks of dust in an evil charade.

Morgellons Disease

Morgellons disease is also called Morgellons Syndrome and some feel that a more accurate description would be "Genetically Modified Organism Disease." I am not an expert on this subject, but I have suffered from a bad case of what appears to be morgellons, since the fall of 2013. Aside from my experiences I have done a bit of research and am doing my best to listen to the heart of my own instincts. Please print and share this article.

Morgellons is a mysterious illnesses, which has been surrounded by misdiagnosis, misinformation and attempts to hide the fact that it is part of a lethal targeting program. Some doctors misdiagnose it as a "delusion" or "mental illness," which merely inflicts more pain upon the victims. Morgellons is so obviously a physical infliction that scientists have launched into research, in order to find its True cause, and have been baffled by the mysterious findings, which show that morgellons Truly IS SOMETHING that is like nothing they have ever seen before. I hope they are soon able to prove and expose its True cause for the sake of victims as well as the future safety of all of humanity. Morgellons, and other parts of the targeting, are in desperate need of being exposed and stopped.

Alien Fibers: Morgellons Disease - ABC's Nightline.
https://youtu.be/xsiJpuARHcE

I'd like to urge honest doctors to gain more trust in their patients and learn how to say, "I don't know," instead of assuming the wrong things, especially when people's lives are at stake. There is not a doubt in my mind that some of the doctors, who leap to the false diagnosis of "delusion" or "mental illness," without even performing any tests, are participants in the program that is targeting us. This same method is used around other parts of the targeting as well.

Morgellons victims seem to know more about the disease then doctors and researchers. I have heard many victims say that morgellons is parasitical. I can attest to this and more.* There is nothing like the wisdom of experience. But, our experiences need to be believed and acknowledged by the medical profession, in order for us to receive needed validation and hope for things to get better for us.

One of the most damaging parts of the covert targeting is the secrecy, and its resulting lack of awareness and disbelief, because this inflicts more pain upon the victims and prevents help that is deperately needed. Most people can deal with almost anything as long as they know what it is and have understanding and support from fellow human beings. Without knowing, and without understanding and support, our suffering is indescribably excruciating and hope for it to end fades into the mysterious dark abyss. One example of a victim who suffers from the mystery is Joni Mitchell. . .

Inside Morgellons: Joni Mitchell's Mystery Illness
https://youtu.be/uBHz5raVMC8

__THE CAUSE__: Morgellons is an infestation of an odd strain of parasite and is a horrible form of torture that is being inflicted upon Targeted Individuals. These parasites are worms that live inside the body and sometimes eat or poke through the skin. They are activated with electromagnetic frequencies being directed into the infested parts of the body. (There may be other types of parasites also.) The worms are probably most often ingested through intentionally infected food or drinks. They can also be surgically placed into the body. . .or remotely shot into the body, which can explain why so many inflictions have been reported to be on the most exposed parts of the victims body - the head, arms and lower legs. But no matter how they get there, the condition is exacerbated with electromagnetic frequencies being directed into the infested parts of the body. The result is uncomfortable, often painful and can be fatal.

These targetings appear to be part of a covert eugenics movement. The hidden technological part of the targeting can be proven with unfiltered microwave detection and radiation tests...etc. And I pray that we soon have enough professionals, who have the Heart and courage to stand up for humanity in public proclamations, because the confusion, suffering and cruel destruction of human life should not be allowed to continue and grow.

THE CURE: Morgellons can be completely cured through the sadistic targeting of human beings being exposed and stopped. Until this happens we can use herbal cures. **Large doses of garlic have helped me a lot**. My herb book listed things like Black Walnut and Chaparral being good for parasites. Some victims say that they have been helped with apple cider vinegar. But, even if the herbs can completely cure morgellons, the targeting needs to be stoped, in order for us to remain healthy.

A POSSIBLE DANGER: Morgellons being a new illness, which is being inflicted upon Targeted Individuals, leads me to wonder if there may be a plan to hide the technological part of targeting through blaming our symptoms on the parasites. It would fit the targeting patterns for the technological parts (especially the technological torture and mind control parts) to be blamed on something else. . .so that it can secretly continue.

A few of the symptoms of BOTH morgellons and microwave targeting are "difficulty paying attention and concentrating, extreme fatigue, hair loss, joint and muscle pain, nervous system problems, tooth loss, sleep problems and short-term memory loss." And there is a desperate need for ALL of the targeting to be exposed, especially the technological parts, so that it can be genuinely stopped for victims. . .and the rest of humanity can remain safe.

This next video contains the most accurate description of morgellons that I have found. In it a woman explains how she did a 4 day detox, using water infused with silver, and found small mushy worms escaping her body. I have experienced similar results while taking large doses of garlic. But I am not sure about the silver water. As I started taking it I felt unsure of what other effects it may have on me. Because I do not know enough about it, I decided to stop and continue with only the garlic and herbs.

Connie Tells Her Story of Triumph Against Morgellons
https://youtu.be/XcL8HvA7sME

The following video is about a woman who believes that she was cured from morgellons through replacing fillings in her teeth. About ten years ago I read a similar story about lupus. I believe that both lupus and morgellons are being caused by microwave targeting, which can be stopped at strategic times, in order to make it look like something else was causing it. This seems to be a common pattern in the aims to prevent people from realizing that they are being targeted with microwave and laser weapons. . .which can inflict a variety of skin problems and physical sensations aside from morgellons.

Perhaps this woman did not really have morgellons. An allergic reaction to a leaking tooth filling is not morgellons. Its an alergic reaction. But perhaps she was being targeted.

Its hard to tell without the technological targeting being acknowledged and the proper tests performed.

Morgellons Treatment ABC interview with Dr. Omar Amin
https://youtu.be/6rJqLnCaPHs

* I experienced small cuts in the skin, around my vagina, before the mergellons got really bad in that area. It appears that the worms were either put into me while I was asleep or the cuts were remotely inflicted with laser weapons just to make me think they were. I can not be 100 percent sure of how they got there, but I am sure they are there, because I have seen them and have the same types of open bleeding sores that are shown in pictures of morgellons victims.

I also get rounds of sores on my head. It has felt like something was shot into my cheek, which created an unexplainable open sore. I have also experienced this in the lower part of my ear.

My experiences with the radio wave (microwave) part of the targeting leaves no room for doubt of what is happening to me. The proof is in the timing of attacks and what I feel in my body. I often feel the faint vibration of microwaves being directed into the middle part of my body BEFORE the crawling sensations and intense itching starts. This type of targeting has often been done to me directly after I lay down to sleep. And it has stopped, when I suddenly flip over and move to a different location on the mattress, too often to be called a coincidence.

I also often experience the same sort of microwave vibrations, which causes intense itching, around my shoulder blades. (I do not have morgellons there.) This is one of the milder forms of the torture, which just creates discomfort and the irritation of repeatedly experiencing an itch in a place that either can not be reached or a place that is embarrassing to scratch in public.

I cannot help but wonder. . . If I cure myself of morgellons, will they try to pretend that I was never really being targeted and continue with the less obvious parts of it? I guess this could work if its all going to be blamed on morgellons. But the Truth is that I was being very obviously targeted, with microwaves, laser weapons and psychotronic weapons, long before being inflicted with morgellons. I know for a fact that morgellons has not been the core of my problem. Its just a gruesome part of the targeting. And I need it ALL to be stopped.

Since I started writing and perfecting this article I am receiving another round of covert threats against myself and estranged family members, which appears to include the infliction of alzheimer's and the death of my father.

God help us all.

Help for Targeted Individuals

If you are a Targeted Individual who is searching the web for help you are probably being surrounded by confusing misinformation. Unfortunately there is not much authentic help on web and most of us fight this battle alone. BUT THERE IS HOPE. YOU ARE NOT ALONE. WE STAND ALONE. . .TOGETHER.

Message for Targeted Individuals
https://www.youtube.com/watch?v=VxoBeP7L5mc

*Imagine a world that understands and cares and helps.
It could happen! It could. It will. Love will win.*

Don't Give up

**Don't ever think we will not escape,
Or that our rescue will remain too late.
Don't give them the power. Don't let them win.
Don't let hope fade. Don't give up again.
Cast aside their darkness and let the Light sing.
Climb up on their stones and raise your broken wing.**

We are worth saving.

I'm sorry I can't put much here, because I am being so heavily targeted that figuring it out has been hard. But here's what I've learned thus far.

Love Yourself - You Do Not Deserve the Targeting
No matter what you have or have not done in your life, you do not deserve to be tortured and harassed. You deserve to be Loved and comforted and protected. Learn to give yourself the care you need, until our fellow human beings are able to be here for us.

BEWARE of the covert "rescue" and other methods of false help; I have experienced those who target me pretending to be help that is trying to covertly rescue me. (They call it "going home.") This "rescue" leads to complete enslavement. This "home" is with the perpetrators. Don't go. Hold onto the faith that someday genuine help with be here for us, openly and honestly. Be Aware of the perpetration pattern of zooming in to be the ones to "help" us in other ways as well. Always take time to think about offers of help and do the best you can to listen to your own instincts above all else.

CAREFUL not to respond to unofficial requests for lists of your witnesses and evidence, because this could lead to them being destroyed.

Listen to your HEART above all else
Heart over Mind is the #1 remedy. I know this can be next to impossible when we are being heavily hit. We are sure to be dooped when we are desperately groping for help and answers while our brains are being lasered and we are fed misinformation. Forgive yourself - forgive your mistakes and just do the best you can. And at least try to remain in a place of compassion for yourself and others instead of fear and anger, it will help preserve our sanity.

Creative visualizations help divert mind control invasions.
TIs should do as much creative visualization as possible. Visualization Example: Sit or lay in a comfortable, relaxed position. Take deep slow breaths. Imagine a protective blanket of pure white light wrapping itself around you. . .and beams of pure white Light streaming down from the heavens and into the top of your head and flowing down through your whole body. Imagine this light washing away all the darkness and radio wave intrusions. Imagine the Light filling you with Peace, protection, strength and Love. Don't forget to keep up the deep, slow, rhythmic breathing. You can enhance this with prayers and also by imagining Angels holding you in their arms. . .or whatever else you find comforting. If you have a hard time fulling visualizing. . .keep at it. It will eventually happen. The more you do this the better you will get at it. Doing some form of creative visualization as many times as you can within EVERY DAY. . .will help more than you can imagine. (Pun intended)

BEWARE of the peaceful music they recommend;
It can be used to help brainwash you. I had a horrible experience with this. Radio, the web and TV are also heavily used for covert messaging. Expose yourself to as little of it as possible.

Spend time in soothing natural environments;
Forests, brooks, lakes and the ocean can help restore your inner balance.

Drink a lot of PURE spring water
The microwaves dehydrate us. But beware of pharmaceuticals in water.

Talk to your Wisdom about not taking unnecessary pharmaceuticals
Mood altering drugs help technological mind control to be successful.

75

Hold tight to Hope and Faith
Your new mantra is "I will be OK." Say it over and over again. Someday, you will be OK. You will.

Let yourself cry and release the pain:
God knows we have a lot to cry about. Its OK to cry. (I just hope you do not get tortured for like I sometimes am.)

Keep a journal of what is happening to you:
Just be careful to not get so caught up in logging perpetration tactics that it consumes you. Sometimes the best protection is to ignore it. Write more about your feelings and less about them. This will help you to process and release some of it. Keep a list of date, time and place of the targeting I have not done a very good job with this, because denial had been my way of coping through many years. Now I find that I have to ignore as much as I can or go completely crazy. However, it does seem important to log as much as possible, without it consuming us, in case help ever arises for us.

Deep breath and remain calm when under attack; Slow deep breaths can help us tolerate the pain and lessen anxiety.

Imagine that you are not alone
This may sound a bit cooky, but please try it. On those days when your heart aches so much that you can hardly stand it. . .and you feel too painfully alone, sit or lay in a comfortable, private place, close your eyes and imagine that an Angel is holding you in hid or her arms. This Angel brings Love and a feeling of peace and comfort. If you are not spiritually based, imagine this Angel being a loved one from your past - someone whom you felt loved and comforted by. Let in that Love. YOU ARE NOT ALONE.

Pray to a Higher Power for help:
I have found it effective to pray out loud for Love and Light to protect me and also flow into those who are attacking me. It doesn't matter what religion you may belong to, if any. Just try to know, in your Heart, that there is goodness in our world, which wants to comfort us and stop the crimes from continuing. Someday it will.

Physical Protection?
I've found that showers sometimes bring temporary relief from microwaves/electromagnetic/psychotronic attacks to my brain.

Lead and other materials can block some radio waves, but targeting can be vamped up to points that make it uncomfortable after it is removed. There seems no physical method of protection unless you are fortunate enough to be able to make or purchase a radio wave blocker. But all modes of protection are only temporary.

The help we need is for criminal use of these technologies to be stopped - for humanity to be set free. The targeting needs to be openly exposed and stopped.

76

!!! TELL PEOPLE ABOUT THIS !!! IT IS THROUGH EDUCATING THE REST OF THE WORLD, OUTSIDE THE TI FORUMS, THAT THE GREATEST HELP CAN BE OBTAINED.

I understand that we are being instructed to remain silent so that we will not be forced into a false mental illness diagnosis. And, unfortunately is a True risk. However, remaining silent offers no chance for help at all. Talk, but just be careful how you say things and try to back it up with things written by other people. PRINT FLYERS AND PASS THEM OUT. LET YOUR HEART LEAD YOU. . .INTO THE HEART OF HUMANITY.

Start a blog or website, using the term "Targeted Individual" in the title and in search terms. In 2012 I had gotten emails, which advised us to not use the "Targeted Individual" term on the web. I believe that this merely hides us and the crimes and prevents us from getting help. Our perpetrators already know who we are and where we are and what we do. Its everyone else, and ourselves, that we need to make a public stand for. So start a blog or website. . .even if all it contains is one page that begs for help and explains a bit about what is happening to you.

Download, print and distribute a free news paper:
http://www.targetedinamerica.com/publicnotice.html

Write letters to government officials

Do not fall into that trap of blaming "the government." They appear to be being targeted too. There seems to be good and bad everywhere in this mess. My gut feeling has been that there is still good in our governments that has just not able to be here for us yet. Letters in every possible avenue for government help is a good thing to do.

Here is an example of one of my statements;
www.targetedinamerica.com/areport11.pdf

Be careful in web forums;

I've heard that some TIs find support on the web. But I have experienced heavy stalking in literally every web forum that I looked for support in. It hurt me more than it helped. I hope your experience is better.

Try not to be too angry with the puppets - don't let them turn you into the lunatic that they often aim to portray us as. It has helped me to realize that the most of web puppets are probably victims of mind control and this lessens my anger and even raises a bit of compassion for them. I hope we ALL regain our Freedom soon. Read the post I wrote on "the recruiting process"; www.sharon-poet-ti.blogspot.com/2014/10/the-infiltration-recruiting-process.html

Use the ORIGINAL Essiac formula

In the spring of 2007 I purchased the four herbs, made the formula, took 2 ounces twice a day and started healing from Lupus, which I believe was caused by radiation from microwave weapons. If you are a Targeted Individual, who is lucky enough to still have your own kitchen, it would greatly benefit you to use

77

this Essiac remedy.

There are many formulas that are called Essiac. So be careful to aim for the original which is reported to be far more effective. Most of the following information is from this site; www.healthfreedom.info/Essiac%20formula.htm

Essiac tea was "inspired by an old Indian medicine man's formula," and is "a proven cure for cancer" as well as being known to produce positive results against lupus - radiation sickness, AIDS, diabetes and almost any other physical ailment. "The original tea combination can be purchased in the USA and Canada through www.Essiac-Tea.org or in the UK through www.cloudstrust.org." But I believe it is better when we make it ourselves. So here is the recipe. . .

6 1/2 cups of burdock root (chopped - about pea size)
1 pound of sheep sorrel herb (powdered with roots included)
1/4 pound of slippery elm bark (powdered)
1 ounce of Turkish rhubarb root (powdered)

Mix these ingredients thoroughly and store in glass jar in dark dry cupboard. The mixture can be steeped to make a mild tea or processed into "the formula" for a stronger concoction. To make the tea: use 1 ounce of herb mixture to 32 ounces of water depending on the amount you want to make.

The original formula, which I found in a book, suggested a ratio of 5 cups of water to one cup of herbs and dosage was 2 oz twice a day and works best when joined with a healthy cleansing diet and a lot of pure clean spring water.

Boil hard for 10 minutes (covered) then turn off heat but leave sitting on warm plate over night (covered). In the morning heat steaming hot and let settle a few minutes, then strain through fine strainer into hot sterilized bottles and let cool. Store in dark cool cupboard. Must be refrigerated when opened.

I can understand that this may not be very consoling to those who are being literally tortured and are desperate for the hell to end. I wish I could give you more hope. I sometimes struggle to hold onto it myself. But I do feel that help will eventually arrive for us and that those who target humanity will be stopped. . .someday.

Read and share my websites;
www.targetedinamerica.com
www.sharonpoet-ti.blogspot.com
http://www.heartbud.com

My writings are sometimes interfered with or altered, so please listen to only the Heart of your own instincts with mine as well as all others.

Don't Give up

Don't ever think we will not escape,
Or that our rescue will remain too late.
Don't give them the power. Don't let them win.
Don't let hope fade. Don't give up again.
Cast aside their darkness and let the Light sing.
Climb up on their stones and raise your broken wing.

How to Help a Targeted Individual

IF YOU'RE NOT A TI AND CAN CARE TO HELP:
We need you to let your heart find the courage
to stand WITH us instead of against us.

Please do not look for faults in us instead of helping us in the ways that are needed. We are surrounded by manipulations that are intentionally set up to make us look bad. The people who target us, and those whom they control, use those set ups to discredit us and make people look down on us. If you look hard enough you are sure to find a new rock to throw at us. Perhaps there is a part of our lives that has not yet been bruised. But I hope you chose to let your Heart refrain from judgment, because we have already been hurt too much. We also desperately need the following things from you.

Open Mindedness: Please do not foolishly ass-u-me that we are "mentally ill," because this adds to our distress at a time when we desperately need the opposite. Misdiagnosis severely adds to our psychological destruction, because it prevents us from getting the types of help that we desperately need and can destroy our trust in our own perceptions. And we should NOT be forced into positions where we have to prove our sanity on top of all else that we are going through! Its just too much. Please open your hearts and minds.

Understanding: Please educate yourselves as much as possible on the subject and practice empathy. We desperately need people to understand and support our fight to survive.

Compassion: Please care enough to listen and not judge, degrade or look down upon us. We need to know that you care.

Stand Up; Surround us with compassionate people who have the courage to make a stand against these crimes. This can offer some protection, because the organized stalkers do not want witnesses. I have begged for this from several aware people and they chose to stand with those who target us instead of me. Fear seems to be the worst enemy right now. And the mind control technologies are used to manipulate those who try to support us. Please let your heart find the strength and courage to bypass the manipulations and STAND UP WITH US.

Public Awareness: Research this and pass the word around ASAP. The more publicity there is, and the more people who stand up and speak out, the better the chances are of ending to these crimes.

Funding: A Targeted Individual who is near the end of the destruction process, is probably left with no home, no family to turn to, physical health problems and

80

not enough resources for protection. (Like I have) This is when a unconditional financial help is deeply needed.

Estranged family members, neighbors, friends and strangers of primary "Targeted Individuals", (especially medical, media, political and law enforcement professionals) are being called to find the Heart, and courage to resist the brainwashings, in order to save our lives and free humanity's future. . .through STANDING with us instead of following direction from those who target us.

I don't want to be left to evil pretenses of helping hands.
I need to be comforted by those who can care to understand.
I don't want to be declared insane for their hateful gain.
I need you to soothe my wounds instead of inflicting more pain.
I don't want you to watch from a silent distance while I die.
I need you here beside me as I pray to God and cry.

P.S. I still stand completely alone in my fight to survive and expose the crimes that are being committed against us. There seems no hope for my family to come around due to the fact that they are long term victims of mind control and some even appear to have been pushed into the program that targets us. I hope other TIs have more success than I have and I hope this info helps to fill their needs. God is here for me, but I hope that someday. . .some human angels are sent my way. I count too. I do!

www.targetedinamerica.com

"The ultimate tragedy is not the oppression and cruelty by the bad people, but the silence over that by the good" ~ Martin Luther King Jr.

~ *My Personal Testimony* ~

I am an unheard victim lost beneath the lies.
I am a tortured one - put on a list to die.
I am a rising wounded - begging for your aide,
Becoming a speck of dust in an evil charade.

But I hope this changes soon

This is just a quick overview. More can be
found in my book, "Targeted in America."
www.targetedinamerica.com

I am a victim of covert targeting, which utilizes satellite surveillance, laser weapons, microwave weapons, psychotronic weapons and chemical warfare, as well as a localized covert harassment program. I am also a witness to the effects of technological targeting on many other people. I've been fighting to publicly expose these crimes, so that help can arrive for us, since 2011, under conditions that are indescribably inhumane.

I've experienced terrifying levels of covert targeting in Canada, Peru, Mexico and in over a dozen states in the USA, including Hawaii, Arizona, Utah, California, Nebraska, Florida, Virginia, Tennessee, North Carolina, Maine, Vermont, New Hampshire, Massachusetts and New York.

I was born in America in 1959. Prior to the targeting yanking the rug out from under my feet I was a mother who lived an ordinary life style, owned my own country homes, was a hard worker who ran my own business and had perfect credit. I had a passion for per-sonal/spiritual growth, writing lyrics/poetry and using herbal remedies. I had/have no crim-inal record and have not engaged in criminal activities. I was certainly not perfect, but was not even close to the kind of person that my government, or anyone else, could even begin to honestly classify as a criminal or a threat to my country or humanity. So I've seri-ously doubted the theory that this targeting is being done by "the government." The core of it feels more like some sort of sadistic ring of organized crime or a dark occult.

The past few years have been a process of my trying to figure out why and how I am being targeted. . .as well as when it all began. I'm still figuring it out. Between rounds of heavy targeting I've been gradually clicking together puzzle pieces that date back to the 1960s and 1970s, and are scattered through my various writings on the web - my blog and my books.

It is nearly impossible to figure it all out, and write perfect reports, WHILE STILL BEING TARGETED WITH RADIO WAVES THAT INTERFERE WITH MY BRAIN FUNCTION...ETC., but I have been doing the best I can. Please understand that possi-ble discrepancies in my writings are a reflection of this grueling process as well as alter-ations by those who infiltrate my computers and web sites.

The two big questions that still echo in my mind are, "When did it all start?" and "Why am I being Targeted?" It has been impossible to be 100% sure of much. The most solid things I have to go on are a few dreams I've had about the targeting and memories of odd things in the past that now look like they were probably part of the targeting. Because of a sister who suddenly had unexplainable nerve damage to her eye, and a brother who suddenly forgot how to read in eighth grade...etc., I now feel that my whole family was probably targeted for the purpose of technological experimentation. . .and that I was later singled out for more heavy targeting. . .perhaps because I was less susceptible to the technological mind control part. The targeting against me started out very inconspicuous

82

with, what now appear to be, obvious vamp ups around 1974, 1977, the mid 1980s, early 1990s, 1999 to 2001...etc. Since 2005 its been like a hell that most people probably could not even imagine.

Like most long term Targeted Individuals I have lost functional ties to loved ones who could have helped me, not only to figure it all out, but to also prove that the targeting really is happening. That vital link to primary witnesses and support were broken before I began realizing that the targeting was happening. Literally everyone, whom I had been close to, has also been targeted, some severely (like me) and most of them with technological mind control to brainwash them into thinking that I am just "mentally ill." Some have been used to help target me or have been recruited into the covert program and used in attempts to "rescue"/enslave me as well.

There are several people, whom I had been close to, who have been being heavily targeted/tortured since the early 1990s, when an obvious vamp up took place. At this point some have been killed and/or "rescued" into the very same program that targets us. I actually do not know if any are still surviving it without being fully controlled by it.

I feel like I stand alone at this point. The one old friend, whom I was able to openly talk to about the targeting, appears to be no longer answering his phone. One of his friends had vanished. I am concerned that the targeting may have vamped up on him since I visited him in 2014. I have been worried about him as well as many others, including myself.

Every day is a struggle for me to survive, both physically and psychologically, especially since I have been isolated from loved ones and financially destroyed. I am now living in a vehicle. But worse than this is happening to me. On almost a daily basis I experience technological intrusions into my brain, which range from mostly mild to moderate with occasional extreme levels of painful torture.

I have experienced weapon attacks to other parts of my body, which have produced pain. . .sometimes in the form of heart attack symptoms, especially when I have done something that they do not approve of, like writing about the targeting. Other attacks have produced burns on my legs and chest.

There is a lot of perversion in the targeting. It appears that microwaves can interfere with digestion, urination, libido, anus muscles...etc. And I have experienced obvious rounds of this since the 1990s. (I also knew other victims who suddenly had the similar complaints starting around the same time.)

For a few years after being drugged and raped anally (in 2005), the weapon attacks had inflicted periods of intense vibrations in that area - like to remind me of the abuse, which I was struggling to deal with.

Since around Nov 2013 I have experienced ongoing attacks to my vaginal area, which appear to include laser weapon attacks, chemical attacks and parasite attacks - perhaps morgalones, or possibly some other form of parasite, since the worst of it started after I found small cuts in my skin. The discomfort appears to be enhanced by microwaves directed at that area. The itching, burning, bruising and bleeding sores...etc., are extremely uncomfortable and often painful. This vaginal attack appears to have started due to my spending two days in a motel room with a man who said he was a TI and had offered to help me after my car had been disabled and I was dumped onto the frigid streets. Those who target me seem to think I did some sort of immoral act with him, although we were NEVER, at any point, physically intimate - no kisses, petting or sex - nothing in that arena at all. We slept in separate beds and I am sure that those who target me know this. But this is not the first time that I have been unjustly judged by those who target me. It appears that if they cannot find a genuine excuse to accuse or attack they can just make up stuff, in order to justify it in their own minds. I guess they think they are punishing me, but its all really just completely unjustifiable cruel, intrusive abuse.

Why do I not seek medical attention? That's another difficult part of the targeting - I have experienced a lot of corruption in the medical field. The targeting has landed me into emergency rooms on many occasions and I have learned that, after my GP died and my medical records were transferred to another doctor, they have been lost and possibly altered. To make a long story short, honest, NON-harmful medical attention cannot be fully counted on for heavily Targeted Individuals like me. . .until the targeting gets acknowledged and exposed and those who participate in it are stopped.

Other forms of targeting, which I have experienced, include covert harassment, which is obviously orchestrated by those who hold me under surveillance. This surveillance includes computerized monitoring of my brain. The "gangstalkers'/puppets are used to walk or drive near me and repeat my words and actions; to inflict verbal degradations and profanity; to perform rounds of loud noises or angry outbursts; to deliver death threats; to help sabotage my relationships, homes and vehicles; to change my phone numbers; to take over old email addresses; to interference with my websites; to spread false rumor campaigns in my communities or work places; to fabricate emails and phone calls; to drug my food, water or surfaces I touch; to drug and rape...etc.

The remote technological mind control part of the targeting seems to be the most difficult for people to believe. But I pray it is realized soon, because it is actually the most lethal part of the long term targeting process. It appears to be happening to far more than just families that are being more heavily targeted. Technological mind control appears to be what makes the rest of the targeting succeed - what makes so many people blindly disbelieve or numbly follow orders to harass me...etc. When I think of the long term effects of technological mind control, I feel scared - not just for myself, my loved ones and citizens of the USA, but also for citizens throughout the globe who are being brainwashed, sometimes even completely controlled, by technologies that are damaging our brains, spirits and souls while preventing the natural process of personal and spiritual growth.

I have experienced four home losses, which I feel were a direct result of the targeting. The home I owned in Loudon, NH was taken by the state of NH DOT through their right of eminent domain. My second home was destroyed in a suspicious fire. I was forced to sell a cabin, which I had owned in New York, due to being followed there by a perpetrator X boyfriend and being targeted by a neighbor who even tried to run me off the road. The Alstead, NH neighborhood I moved to in 2005 was suddenly wiped out in a flash flood, which was caused by an unusual stalled storm and a plugged culvert. (Four of my closest neighbors were killed.) I have been mostly homeless since then.

In the mid to late 1970s I had an odd chain of vehicle accidents, which included one where a vehicle obviously ran me off the road and then took off. Since then I have experienced many rounds of events that appear to be attempts to harm me. But these episodes did not become evident until around 2005 when I started realizing that I was being targeted and began looking back and taking a closer at things that I had previously thought were just a lot of bad luck.

Around 1974, I witnessed some odd behaviors in my patients, and had an odd experience with headaches, while working at the Hillsborough County Nursing Home in Goffstown, NH. The headaches stopped after I quit working there. I was recently told that many long term employees ended up with similar types of cancer. I now believe that this nursing home was being microwaved. . .perhaps also due to technological experimentation. I had initially wondered if the targeting had followed me home from there, but it now appears to have begun before that time. My mother may have been an MKULTRA victim from the Montreal Canada area.

I've had a few dreams that showed something bad happening to me, when I was around 11 years old (in 1970), due to a pale blue coat that was given to me by my fifth grade teacher. Exactly what this could be about is a mystery to me. But I believe that there is something to it.

In the early 1980s I gave birth to two children. One was suddenly breaking out in odd rashes - "heat rashes" the doctor said. (I now believe that it was probably due to microwaves.) My second daughter was born with a minor heart defect and four breasts, which I now believe was caused by the microwave targeting while she was in my womb.

It now appears that, by the early 1990s, my life was being infiltrated by members of what appears to be some sort of dark occult. Their goal appeared to be to take over my home and convert me, and/or gain control over me through coercion and through inflicting emotional pain - surrounding me with discord and one problem or crisis after another.
It appears that mind control technologies and rumor campaigns were used, in order to brainwash some of my loved ones. (I had been close to most of my family and had a lot of good friends and neighbors, prior to the vamp up in the late 1980s and early 1990s.) The manipulations set things up so that I felt like I had to get away from them, in order to stop the chaos and they blamed me for leaving them. Those who could not be turned against me appear to have been heavily targeted and so overwhelmed with their own situations that they can't handle anything else. . .like I am.

I launched into doing my life's work, in the late 1990s, with my second book, "Embracing Feelings." But the finished manuscript was destroyed in a suspicious fire, which raged through my home in 2001. I resurrected it into a book called, "Embracing Sadness" in 2003, but it was never the same.
In 2005 I started a mission through a publication called "the Personal Journal" and was targeted so heavily that I became physically ill and was forced into bankruptcy and into hospital emergency rooms on several occasions. Since then I have struggled to restart my work under different names and in different areas, but with no real success with avoiding the targeting and the sabotaging of my work. Out of these attempts came the "Sharon's Bud" and "Heart Bud" publications along with a few other books. I have been forced to discontinue my sponsored paper publications due to what appears to be serious targeting of those who have sponsored my writings.

By 2006, due to the targeting, I was destitute, had no one whom I could turn to for help or support. I was being microwaved so heavily that I nearly died in the 2006/2007 winter. I began recovering after I moved, changed my name, used the original essiac formula and walked and prayed and cried and wrote through the spring of 2007. This is when I decided to shift my work into a news paper format and wrote the first "Sharon's Bud" publication. But this was heavily targeted as well. My sponsors/advertisers, in the second printing, now appear to have been targeted.
Other jobs I tried to get, in order to survive and resurrect my work, have been quickly, and sometimes painfully, sabotaged. I have recently realized that my writings have been being tampered with - altered or erased since at least 2001. I have experienced a lot of files being erased and dates on files and documents being altered. My computers are often infiltrated, and I sometimes get threats and heavy doses of microwaves and laser shots when I write. . .making my published works rushed and fumbled attempts to continue my work. Over and over again I have been forced to pick up the broken pieces and start over.
In the summer of 2010 I began working at gathering the hope, which had written "Embracing Feelings" in 1999; the inspiration, which started "The Personal Journal" in 2004; the strength that created "Sharon's Bud" in 2007; the courage that wrote "Out of the Dark"/"Into the Light" in 2010; and the Wisdom that is growing from my own mistakes and experiences. . .so that I could begin building them into "The Heart Bud" - into what my life's work was meant to be - a ray of help for the Heart of humanity.

Though I have printed and distributed a few Heart Bud papers, the sabotaging has been so severe, against both myself and my advertisers, that it has not gotten far and is basically at a standstill at this point, except for on the web. At one point a phone company had even continuously changed my phone number, and refused to give it back to me, after I had printed up ads and passed out business cards with that number on it. This happened several times while I was trying to get sponsors for the second Heart Bud printing. This was BEFORE I was writing anything about the targeting.

Since September of 2011 my work became an intense drive to prove and expose the targeting so that freedom is regained, lives can be saved and I can get back to my work.

Some people think that the targeting would stop if I stop writing about it, but this is not so and it feels like a foolish way to blame me for what criminals are doing. In some ways, the targeting was worse before I started writing about it. Exposing the targeting has made parts of it back off to some degree, although it has vamped up in other ways.

The rounds of chaos and difficulties, instigated by the targeting had followed me everywhere I went and remained so covert that I did not even realize that I was actually being targeted until the end of 2005. Prior to that I could not even imagine that things like this were happening to people (especially in the USA) - I thought I was just having a lot of bad luck and looked upon the difficulties as a spiritual challenge - as opportunities to feel and heal from what was happening to me. It also lead me to believe that the Heart of humanity was being lost. . .little did I know at that time just how devastatingly true this is.

My "spiritual" outlook is what inspired most of my publications.
www.poeticpublications.com

What I've been experiencing, since 2005, is so cruel and so horrible that I sometimes wonder how I'm surviving it. The pain that I, and those whom I've been closest to, have been inflicted with has extended so far beyond excruciating that it has been impossible to fully process while being targeted.

My life has become an intense struggle to survive while I am being almost continuously either microwaved, lasered, gripped with psychotronic weapons, stalked, harassed and sometimes threatened, drugged and attacked with chemicals or bacteria...etc. And the obvious aims to discredit me, through framing me for crimes and/or mental illness, have been intense to say the least.

It has been difficult to effectively write about the targeting on my blogs and websites, while being so heavily targeted. When I look at the scope of what I've been through in the past few years, its a miracle that I am doing anything at all. I continue out of desperation for these crimes to be exposed and stopped, because I have seen too many people being too severely hurt, myself included. Our suffering is immense. I feel trapped and in desperate need of a miracle that can provide me with the protection and financial help that I deeply need, in order to even just start recovering from it all.

But the saga relentlessly continues year after year and I feel like I'm not fully surviving it, at this point. I am getting worn down by round after round of vamped up targetings. However, I have deep faith in God/Love/Light, which is carrying me through. **God, help us all - help our Freedom to be regained.**

Report of Most Obvious Targeting

I was born in the USA and have resided here all my life. It appears that I've been being targeted since at least 1974, although I did not start realizing it until 2005. I am now experiencing rounds of heavy targeting, which include technological torture, covert harassment, the sabotaging of my relationships, homes and work and attempts to harm me. **Many people, whom I have been close to, have also been targeted in various ways. Some have been literally enslaved. I am deeply concerned for their lives and psychological health as well as my own**.

I have experienced covert harassment and microwave weapon attacks in Peru, Canada and many

86

states in the USA, including New Hampshire, Maine, Vermont, Arizona, Utah, Virginia, California...etc. **Since 2006 I have been sending a variety of reports to every level of law enforcement in the USA, as well as some internationally. Many of those, whom I report the targeting to, appear to be either targeted or prevented from helping in other ways. I feel scared for all of us, especially for individuals and families that are being harmed. Please help us. We need protection from further harm.**

This list includes just a few examples of my most provable experiences. More can be found on my web-site and in PDFs of the books listed below. **Please forward this to appropriate officials.**

Web Site: www.targetedinamerica.com
"Targeted in America" book; www.poeticpublications.com/booktia5.pdf
"Ramblings of a Targeted Individual" book; www.poeticpublications.com/bookram13.pdf

Sabotaging of my Work and Targeting of My Sponsors

My work has been sabotaged in ways, which include interference with mailings, the destruction of my computers and printers, Effecting my health with microwave and laser weapons, infiltrations into my com-puters to alter or plagiarize my writings, changing my phone numbers, taking over my email address, wip-ing pages off my web sites...etc. But the worst of it has been the targeting of my advertisers since the 2008 printing of Sharon's Bud.

In 2014 Daniel Nadeau promised to sponsor my next Heart Bud publication shortly before his "Gold Buyers" shop in Greenland, NH was broken into. This was my last attempt to find sponsors for my publica-tions, because **there has been a long chain of what appears to be targetings of my sponsors in the second issue of Sharon's Bud and four issues of the Heart Bud.** (Possibly even subscribers of my 2005 "Personal Journal" publications.) **Most (if not all) of my advertisers appear to have either gone out of business or been targeted with break-ins, bursting pipes, sudden odd deaths or illnesses of loved ones...etc.** So much has happened that I can no longer see it as a coincidence.

Since I learned of Dan's suspicious "death," (which is listed below) I began to realize that things appear to be getting worse instead of better in my situation, especially for past advertisers of my papers, witnesses to the targeting, and possibly even people whom I had written about in the original posts on my blog, which are now in the book listed below . . .or should be. There are things that may have been altered or erased by those who infiltrate my computers and web sites. (I've been getting what appears to be covert threats connected to the writings in this book.) I feel bad that I must share it in its raw form, although it has been being infiltrated and I had made too many of my own mistakes, misperceptions...etc., and certainly do not want to misinform or let blame aim in the wrong directions. But I can not properly fix/edit it on infiltrated computers and feel that I need to share it so that the crimes do not remain hidden. **More complete ear-lier editions can be made available to the proper authorities upon request.** Hopefully they will remain in tact.

The "Sharon's Bud" and Heart Bud Publications; www.heartbud.com
Ramblings of a Targeted Individual Blog; www.sharonpoet-ti.blogspot.com
Ramblings of a Targeted Individual Book; www.poeticpublications.com/bookram13.pdf
(More complete pervious editions can be made available upon request.)

Since my work has been so severely sabotaged I have repeatedly aimed for other jobs. On those jobs I've experienced what appears to be laser weapon attacks to my back, hands and arms, which were inflicted to disabling me until the job is lost or not obtained. On jobs that I have obtained I've experienced severe levels of harassment (including sexual), rumor campaigns, being drugged and raped, microwave weapon attacks to make me physically ill or interfere with my brain function.

Suspicious Deaths

Before you read this I want you to know that I do not have an issue with death. I fully believe that our souls/spirits go to a better place after our bodies die and I actually find comfort in knowing this. But there are some deaths surrounding my situation, which feel too unnatural and appear to be inconspicuous mur-ders that are set up to look like suicides or natural deaths. Some appear to be part of a targeting and abduction/"rescue" process that is orchestrated by those who target and enslave people. Some enslaved victims are thought to be dead by loved ones - their "deaths" staged. I beg you to help stop these crimes and expose the covert program so that enslaved victims can be set free and not used to "recruit" or target the rest of us.

Death of Daniel Nadeau of Greenland, NH in 2015:

Dan had been one of my advertisers in the 2008 second printing of Sharon's Bud. Dan had also helped me when my car was disabled and I was left stranded on the streets in the fall of 2013. On February 28, 2014 I reported that Dan's "Gold Buyers" shop was broken into shortly after he had agreed to sponsor the next printing of my "Heart Bud" publication. . .and that I thought it was part of the targeting.

Around July 28, 2015 it appeared that Dan was being used to try to get me to leap into his car - to "rescue" me from the targeting. **"He won't go without you,"** one of them said as this was happening. On one occasion Dan parked next to me, walked in front of my parked car and then appeared to be sitting in his car waiting for me to leap into the back seat. Those who target me have recruited and/or abducted and used, many people whom I knew or had been close to, and use them to try to covertly "rescue"/abduct me. Due to my experiences with this, I feel that this "rescue" (into a place they call "home") leads to complete enslavement under the very same people who are targeting all of us.

On September 6, 2015 I heard that Dan was reported to have died of a heart attack on February 22, 2015, months BEFORE I saw him. Is he really dead? Cover-ups are happening around this. The dates appear to have been altered in my reports of this. And my February 28, 2014 blog post had been partially erased after his death.

Dan is one of several people, whom I now feel are probably still alive and enslaved although thought to be dead by loved ones and the rest of the world. It appears that staged deaths, which hide abductions, are merely one of the recruiting methods for those who target and enslave people.

Four Deaths in Northfield, Vermont in 2012:

I visited a church in Northfield and befriended the minister and a few other members of the congregation as I sold books and looked for a room rental in this area where I had just obtained a job. (none of these books were about the targeting.) Within just a few weeks that community, which had embraced me with kind, open arms, was stricken with four sudden deaths. One was the elderly man whom I had breakfast with during my first visit. He was missing on the day I returned and was later found dead in his field. The cause was reported to be a heart attack. This was followed by three other deaths, which had the whole community spinning in every direction but mine. This was one of the more extreme situations where people seemed to be being targeted, and distracted by deaths of loved ones, due to their care for me. This sort of thing has happened often, probably more than I realize or even want to know right now.

Death of the two aunts who would have helped me:

In the summer of 2008 I sent a letter to my aunt Alene Coache, who was a Nun in Montreal Canada. (I was wanting to go see her about the targeting) After a period of no response, I found out, through the internet, that she had died of a sudden illness. This same thing also happened in 2006, when I was about to go see my aunt Francis Rowland in Florida. Her first husband (my uncle) worked for the FBI and I there may be a connection between his job and our family being targeted.

The Psychological Death of a Child in late 1990s and the following decade:

I worked at a mental health facility for a few years and worked with a two year old boy. He was having adverse reactions to psychiatric drugs. I reported this to the lead psychiatrist who kept trying different ones. I apposed it, because he was normal - nothing wrong with him. But his psychiatrist said that keeping him medicated and quiet was helping his mother to treat him better. I was taken off the case and , around a decade later, I was told that that little boy had become retarded and was institutionalized. I am deeply concerned that these sorts of things are happening to many people.

Kevin LaBree's Death on River Road in Goffstown or New Boston, NH in 2002:

In Dec 2001 my brother Kevin LaBree called me and said, **"Something weird is going on around here! Pops is suddenly acting strange..."** ("Pops" is what he called our father.) I did not know what to think of it and was being so heavily targeted that I forgot it. But it is now clear to me that he was starting to realize the technological mind control part of the targeting, which none of us were aware of at that time.

In August 2002 Kevin was reported to have been killed in a mysterious accident with a four wheeler and was found face down in a river in New Boston, or Goffstown, NH. Many people knew that there was something suspicious about his "accident" but, as they pushed for investigations, my cousin suddenly died of a heart attack. I believe that the second death was a murder set up as a distraction, because this is a typical pattern of the targeting. I now have reason to believe that Kevin is still alive - that his death was also staged and that he is enslaved in the covert program. (His body was not cremated and is supposedly buried in the family graveyard in New Boston, NH.)

Within two days BEFORE my brother's "death", a perpetrator puppet had talked about being scared of his little brother dieing. And this is a classic type of sadistic forewarning that those who target me put out before executing their plans.

Death of Jim Baker in Nova Scotia, Canada in 2001:

Around a serious vamp up in the targeting, and shortly before a suspicious fire destroyed my Andover (Potter Place) NH home, a friend of mine, who was selling me a piece of land in Nova Scotia, was reported to have been found dead in his house near Berwick, Nova Scotia. This was reported to be from a heart attack. But I have reason to believe that he is still alive - that he is yet another person who has been abducted/enslaved.

My oldest daughter was surrounded by an unusual amount of deaths:

Around the late 1990s to 2001...etc., My daughter's friend's grandmother, whom she'd been close to, passed away and she had a hard time dealing with it. . .and then she was suddenly surrounded by unusual deaths. (This is a pattern in the targeting - instigating things that we have issues with.)

Another friend's mother was suddenly found dead with both of her grandchildren, in Penacook, NH. It was reported that she had killed her grandchildren and then herself. (I do not believe that this woman did such a thing on her own accord.) One of her other close friends got married and then her husband shot himself in front of her. They called it suicide. (around this time we also lost three pets.) A few years later she walked into her co-worker's apartment and found him dead. Reports said it was from diabetes. (Both of my daughters have been targeted/traumatized in various ways.)

Death of Charlie Buck of Mont Vernon, NH in late 1980s or early 1990s:

I have reason to believe that my father in law, Charlie Buck was targeted into inhumane suffering while being murdered with microwave weapons. He truly is in a better place now, but how many other people are being harmed the way he was?

Death of my mother, Yolande LaBree, in 1977:

I now have reason to believe that my mother never died - that her death was also staged and that she was forced into covert enslavement under the guise of a covert "rescue" from a hospital in New Hampshire.

Examples of Attempts to Harm Me

These are just a few of the more obvious things. I may never be able to prove them, due to efforts to hide the crimes, especially in the past three years, but they did indeed happen. Since the end of 2005 I have been literally fearing for my life, and that of estranged loved ones...etc. Please help us.

Metal Object Shot at my Car in Kittery, Maine in Summer of 2015:

This hit my windshield just a short distance from my open window and head and was done after a threat was delivered.

Shot With a Laser Weapon in Portsmouth, NH on June of 2013:

This gave me severe heart attack symptoms with a laser type of weapon that delivered an intense vibration aimed at my chest. When I moved I felt the vibration in my neck and head. My heart reacted violently. I experienced severe pain in my chest and could hardly breath for a while. It took a few days to completely recover.

Inflicted With Anthrax in York Beach, Maine in 2009 or 2010:

In November 2009, I rented a room in a rooming house in York Beach, Maine. Between the time of my looking at it and deciding to rent it, the manager told me that a couple of Navy boys had just called to rent two rooms after I left.

On the evening of December 23rd my lungs had a strange reaction to something directly after I climbed into bed. On December 24, 2009 I worked through the day while enduring the worse type pain I'd ever felt in my lungs. Later that night I was sitting in the kitchen when one of the Navy boys came in, took a phone call and then rushed out the back door. I suddenly felt like something was wrong. I rushed to throw my belongings into my car, but was hit again, while in that process. I went to the hospital but they seemed to think I was just crazy, because I said I was being targeted.

Through the next few days I remained in a lot of pain, had a hard time breathing and continued coughing up globs of mucus that had little back dots in it. Someone told me that it sounded like symptoms of anthrax. But as I went to another clinic to be tested they referred me to a government agency who told me that there had just been an anthrax exposure reported and they were suddenly flooded with phone calls, from people who feared possible anthrax exposure, and mine got lost in that shuffle. This sudden anthrax scare, right at the time when I was trying to be tested for it, feels like too much of a coincidence and is typical of the distractions set up to cover the targeting.

As I recall this chain of events, now that I am a lot more aware of the tactics of those who target me, my mind is filling with questions; Why would the Navy send people after me, and if they did, would they foolishly make it so obvious? Were those guys REALLY from the Navy or did someone just want me to think they were? I am recently realizing that the more lethal parts of the targeting are often surrounded by manipulations designed to make us blame the wrong people. I now feel that this was probably one of them.

Loss of Brakes Near Needles California in January 2009:

On highway 40 I suddenly experienced the loss of my brakes and what appeared to be the microwaving of my vehicle. The whole electrical system went down and stopped the vehicle, which started rolling backwards toward a cliff. I shoved a block of wood under the wheel and called 911 for help that never arrived. This was after I had tried to report being drugged and raped at a driving job and had sent a sample of drugged water to the LA FBI. . .and after visiting a woman whom I now realize was connected to a group who was part of the targeting. I had the truck towed and a garage attendant told me that BOTH rear brakes had been "over adjusted" causing them to fall apart and malfunction. Oddly, he had me go in after hours, in order to help me through telling me that someone was trying to harm me. He said that, if he had told me as an employee of this garage he'd have been fired. I found this odd. But I am now concerned that worse may have happened to him since I have realized that I am held under surveillance by those who target me. They probably know that he helped me and was an important witness.

Attempted Murder on Saint Lawrence River in Alexandria Bay, NY in 2006;

In 2006, I was aiming to photograph Light Houses for a series of gift cards and Poetography prints, which I was creating for my work. I'd shared this with people and was asking about how to get to a light house, when a man told me that some could only be viewed from the river. He offered to take me out on his boat to a few Light Houses on the Saint Lawrence river in Alexandria Bay, New York, which is where I was at that time. I took him up on the offer and got two great pictures. But something odd happened out in the middle of the river he turned to me and asked if I could hand him his jacket (or something) at the back of the boat. I stood up and went to the back of the small boat and then he suddenly full throdled it. . .nearly knocking me overboard. Thank God I was able to grab something, which prevented it. He behaved strangely (very quiet and hardly looked at me) after that. Oddly, at the time, I did not even think that he had done this intentionally. I'd thought he had probably slipped and fell into the throdle. . .or something like that. But I now feel that it was an intentional attempt to harm me. . .possibly even kill me.

This happened within a month after I had told someone that I was drugged and raped by a couple members of the occult I was being targeted by - it happened as I was fully facing what had happened to me and as I was aiming in the direction of reporting it.

Diagnosed with Lupus in 2006 in New York;

After I reported the rape and after the attempt to make me fall off a boat, I suddenly got really sick and went to the emergency room. A doctor too blood tests and said that my SED rate was dangerously high - that I was near death. Another doctor ruled out other possible caused with new tests and said that I had lupus. (I now feel sure that this was microwave induced.) Shortly before January 18, 2006 I publicly declared that the Lupus, which I'd just been told I had, was caused by harmful energy that was being directed at me from a dark occult. This conclusion was partly due to sets of dreams I'd had, which appeared to be warning me that I was sick due to harmful energy being directed at me. (At that time I was not aware of the microwave weapons, which I now feel it was being done with.) After publicly sharing this I was attacked so severely that it nearly put me back in the hospital. Around this time my business email address had been taken over and I was being heavily lasered while driving, so badly that they appeared to be trying to make me get into an accident, especially as I aimed to leave the area - Clayton, NY.

Parachuting Accident in 1999 or 2000;

I went parachuting with someone. The instructor, whom I was lined up to jump with, suddenly decided to take someone else before me, because that person was in a rush. Then the instructor told me that, on that jump, which I was lined up for, his parachute failed and he ended up having to cut it off and open an emergency back up shoot. He was so shaken, he said, that he was not doing another jump that day. I never got my turn, which is probably a good thing for both of us.

Attempted Murder in Peru in 1999;

While hiking the Inca trail the leader of my group was overly persistent with wanting to help me

90

over dangerous parts of the trail. Aside from him I was the most experienced hiker in the group. I kept refusing his help, but he persisted. During the one time when I relented and took his hand, to round a dangerous drop off, on the tall mountain that juts up above Machu Pichu, he suddenly let go of my hand - pushing it outward as he appeared to have tripped. Luckily I'd still had my left hand on a rope and had not been fully relying on him. The drop would have surely killed me. I now feel that his "tripping" at that moment was not accidental. This trip was suggested by a friend of his - the woman whom I had just told about the dreams I had about something bad being put into public water supplies and about her taking bad water to children at a school.

That first trip was so bad that I vowed to return to Peru by myself. This second trip was in **September 2001**. And to make a long story short, it was filled with so many odd occurrences that they can NOT be viewed as coincidence. On the day I was to return to Cusco from Machu Pichu, the train suddenly broke down and I was flown out in the cargo part of an old helicopter, where the stench of gas was almost too much to bear and left me feeling sick. As I aimed to fly from Cusco to Lima, on my way back to America, the flight was suddenly canceled or late and I was rerouted onto another flight, which I was told was the last one to leave due to it reaching a time of day when it was too dangerous to fly over the mountain range that I was rerouted over, in order to catch another flight that would then take me to Lima. The flight was terrifying for all passengers. The plane kept dropping out of the air. We thought we were going to crash on a number of occasions. People were screaming and vomiting. It was horrible. Oddly, the attendant, instead of apologizing to the passengers, had smiled and said, "I hope you enjoyed your flight" and some passengers became enraged and yelled at her.

Dumped beneath dangerous rapids in an ice cold river in Andover, NH in late 1990s;

A man had tried to harm me while we were white water rafting. The raft was stuck at the base of a waterfall and he stepped onto a rock, climbed out and tipped it over, dumping me under a powerful water fall that held me under the water. Then he kept pulling the raft away from me as I surfaced choking and groping for it. I nearly drowned - I inhaled a lot of water. We were rescue from an island by one of my neighbors, because I refused to get back in the raft with him. On another occasion he had thrown a 4x6 timber at me, through a window in my home. On another occasion he'd tried to inconspicuously shove my hand into a running saw blade by pushing on a board I was about to cut, while pretending that he we was just trying help me by adjusting it. This man was obvious operative for those who target me. I was in a relationship with him for about 3 years, during which time I now feel that he was drugging me and harming my daughters as well. (He was also the one who later told me that he was concerned about his little brother dieing just before my little brother died.)

Unexplainable Loss of Head Lights on my car in NH in the late 1980s and early 1990s;

The head lights on my car would suddenly stop working while I was driving in the dark. Mechanics couldn't find anything wrong with it. This never happened when my husband drove it. But it kept happening to me. I now feel that this was being done remotely and to stop me from visiting my step father, whom I now feel certain was being murdered with microwaves. Oddly, I can not remember the year but it was in the late 1980s to early 1990s.

The Mill Fire in New Boston, NH in 1980;

In the spring of 1980 I was between 7 and 9 months pregnant, and at my father's home baby sitting my twin niece and nephew, when a fire suddenly broke out in my father's shop. I saw the smoke through the kitchen window and rushed upstairs to get my little brother. (We were the only ones there - my father had gone shopping.) While my brother raced outside I picked up the phone, but it was dead. (There was no reason for the phone to have not been working - the fire was not near the house or power lines.) I was unable to call for help so I rushed outside, where my brother was trying to access the building and move my father's trucks. I yelled for him to take my car to a neighbor's house and call the fire department, because I was caring for my little niece and nephew.

He raced off in my cougar and a few minutes later, our closest neighbor drove into the yard with two other men crammed into the cab of his truck. "You'd better move that gas tank or the house will blow up with it," he called out as he turned around and all three of them drove away, leaving me there, obviously pregnant and with two scared and restless 4 year old children. Fear gripped me at this point. The flames were circling around a diesel fuel tank and getting closer to the larger gas tank he'd referred to. I rushed the children to the other side of the house, in case it exploded. And every minute felt like an hour while I waited for help to arrive. There was no way that I could move the gas tank by myself even if I did not have to take car of the children.

Meanwhile, a tire fell off of my car while my brother was racing for help. (The lug nuts had been unscrewed.) But someone happened by and the fire department was eventually called and arrived after the building was mostly burned down. This was an old farm building that was about 200 feet long and 50 feet wide.

The cause of the fire was never determined. Nobody, that we knew of, had been in the building for about an hour, and so it was assumed that it may have been caused by a smoldering spark from a grinder that my older brother had been using before he left.

Rumors spread that my father torched it for insurance money, which added to the distress. He'd have never done such a thing. It was not even insured. I have recently realized that the false rumor is a typical pattern of the type of targeting I've been experiencing - the exact same rumor was spread when my home was destroyed in a suspicious fire almost two decades later.

Now that I know we are being targeted, its like another puzzle piece clicking into place - it appears that the fire was set and that the phone and vehicle were disabled, in order to prevent help from arriving. (This property was a large farm and was 4 miles from the center of a small town.)

Motorcycle Accident in Goffstown, NH in 1977;

In the 1977 summer, while my mother was in the hospital, an elderly man pulled out in front of me. My bike, struck his car sending me flying over it and landing hard on the pavement where I remained, in horrible pain, until an ambulance came. The arrival of the ambulance ended up being an extended period of time, because, one of the EMTs suddenly started having heart attack symptoms and the driver was forced to turn around and bring him to the hospital, instead of picking me up. (This was on the front page of that weeks local news paper with a picture of me being hauled off on a stretcher.) Meanwhile it had started to rain and a couple of police officers were holding a tarp over me until a second ambulance could get to me.

Another unusual thing about this accident is the way someone, who had witnessed the accident, had run out yelling and screaming at the elderly man who'd pulled out in front of me. "Look what you have done! You've killed her!" he screamed. Now, in looking back at it, it seems like this was staged.

I was run off the road on route 101 in New Hampshire in 1975 or 1976;

I ended up in a serious car crash that destroyed my car, broke my collar bone and gave me a concussion. There was a witness to this. I must have gone unconscious for a little while and, when I came to, a police officer was standing at the door of my car and a woman was crying out, "they just ran her right off the road!"

I was taken to the hospital and tortured by a doctor who had two orderlys pin me down, without ever explaining what they were doing, while the doctor shoved my collar bone back into place. (This same doctor also tortured my little brother through stitching his head with no novocaine...etc.)

My mother must have sensed something wrong, because she came to the hospital and took me home, against doctors orders, while I was so drugged that the details are a bit foggy. Shortly after this, my mother was diagnosed with leukemia and we lost her just before Christmas in 1977.

Sudden loss of brakes in Goffstown, NH, In 1976;

I approached a stop sign, went to hit the brakes and the pedal went to the floor - I had no brakes! I flew right through the stop sign at the corner of route 13 and Bog Road and almost hit a truck, and then plowed into a snow bank in order to stop the car.

Exactly why the brake fluid had suddenly drained out of my car was a mystery and the mechanic/friend who had worked on the car, prior to the brake loss, assumed that maybe he had absent mindedly left a wrench on the bleeder and it had loosened the bolt as the car bounced over bumpy roads. None of us had even considered the possibility that someone may have intentionally drained the brake fluid out of my car, because I had no idea that I was being targeted or that anyone would want to harm me.

But I now believe that this is most likely what had happened. I have learned that the targeting is often set up to place blame on someone else. Tampering with my brakes after a mechanic had just worked on them would be the type of timing they'd leap to take advantage of.

The Targeting of My Homes

The most obvious targeting events have included the taking or destruction of my homes. I may not be able to prove most of the details, especially if things are covered up and witnesses are targeted into silence one way or the other. And unfortunately it looks like this has already been happening to some degree. But my statements below are true.

The NH Department of Transportation Taking of my Loudon, NH home in 1995;

There appeared to be repeated attempts for perpetrators to take over my country home, on the corner of route 106 and Staniels Road in Loudon, NH, after my husband and I divorced. Two people offered to buy it and let me continue living there. Two had tried to actually move in with me. All of it was done under the guise of help and by those whom I am now realizing were part of a perpetration infiltration into my life in the early 1990s. I had refused all of their offers. And then the DOT took my home under their rights of eminent domain.

The process of the DOT taking my home was dragged out. . .leaving me in an uncomfortable position, financially, because their plans to put a road through my property were preventing me from selling a commercial part of it and from freely continuing my in home business.

At one point I had called a news paper reporter with the hope that some exposure would help swing things into a more positive direction. But the reporter stopped by the DOT office before coming to my house and the DOT called me while he was driving from there to meet me, and informed me that everything was looking better and set the date for the closing. I didn't realize that this was a manipulation at the time. I had believed that things were genuinely swinging onto a better course and that there was no point in pointing out the negative details to the reporter. When the reporter got to my house, I told him that the DOT had just called and that everything had changed. "I'll bet they did," he said. The phone call from the DOT and my lack of awareness set the course for a news paper article in the Manchester Union Leader, which made it all appear like it was a good thing. But it wasn't and things got worse. . .and it seemed like there was nothing I could do to stop any of it. Local lawyers even refused to help me.

The closure date was delayed after I'd shut down my business in preparation for the move. This left me with almost no income. An official at the DOT suggested that I go on welfare. I was mortified. (The aim to sabotage my homes and work and try to push me into welfare or disability is a strong pattern in the targeting.)

I remember getting a call from a DOT official after the news paper article came out. He told me that it would be best if I did not talk to anymore reporters. And then he said that he was friends with the president of the bank who held my mortgage and was having dinner with him. After this he said something like, "Do you know what I mean?" But I didn't know what he meant, at the time. I thought it meant that he was going to put in a good word for me. I followed through with my plan to visit the president of my bank, explained the situation to him and asked if I could make lower payments - just the principle part of the mortgage, until the state followed through with the purchase and it could all be paid off. Within a couple days a really mean sounding thug called and said that my home would be immediately foreclosed on if I were to be late on just one payment. This was shocking, because I had NEVER been late on a payment and had perfect credit. I had gone to talk to them, with the hope of preventing such problems.

A friend's father had stepped in to help me. I think he had called the DOT and then he come to the closing, which is probably what made things go better than they had been. (But that friend's family appears to have been targeted as well.)

During the taking of my home I rented a room to a woman whom I now believe may have been drugging me. My reactions to the DOT became extremely uncharacteristic of me. At one point I wrote F----- You in a letter to them. This word was not even in my vocabulary! And I had changed my name from Sharon Buck to Namatari Neachi in the middle of the process, which is also odd.

The Fire in My Andover (Potter Place), New Hampshire Home in 2001:

I am in the process of remembering more of the details, which surrounded the suspicious fire, which destroyed my Andover (Potter Place) New Hampshire home in 2001. Like other heavy parts of the targeting, it was surrounded by extenuating circumstances and odd chains of deaths and events, but this is just about the fire. Around the evening of May 7, 2001 I arrived home from shopping and my house was on fire. I called the fire department. But there were a few odd things that surrounded the fire and the process of putting it out.

1. I heard that there were two other fires on the same night. And I think that the first responding fire department, which took charge at my house, was not the local one. As things seemed to drag on with what appeared to be no attempt to put out the fire, a local fireman, who was off duty, tried to find out what was happening. I remember him apologizing to me and frustratedly saying, "I'm not in charge here." Apparently they sent him away.

It was hours before a pool of water was set up in the yard to extinguish the flames. The odd thing about this was that I had a large brook in my back yard, which the local fire department had used for a practice drill, to help me fill my swimming pool with water, just a few years before this. The water was already there and this was well known by the the local fire department.

2. After the fire was finally out, and the fire trucks were pulling out, a police officer came to me and asked if I wanted him to call the fire marshal and have an investigation started. As I thought about it he said, "I would if I were you." And I agreed.

93

3. The fire marshal, and a few other people investigated for a few days. The marshal told me that the burn pattern was suspicious - that due to his findings, and a police report, it looked like the fire was put out in one room and then restarted in another room.

The fire marshal was trying to figure out how the fire jumped from one room to the other without the normal burn pattern. He'd repeatedly asked me what was in the room, where the fire appeared to have been restarted, but my mind kept going completely blank; for some odd reason, I could not remember what was in that room, while they were there and questioning me. It ended up being listed as an unknown cause.

I later remembered what was in the room, because it was the room, which contained my most cherished personal belongings, and I soon started missing them. It contained my journals, where I had logged nearly three decades of experiences and dreams; notebooks filled with over three decades of my poetry; the final manuscript to a book, which I'd just written on the subject of "Embracing Feelings" and avoiding psychiatric pharmaceuticals; thousands of dollars in cash; my address book - all of my personal contacts; my clothes...etc. (Please read this article; www.targetedinamerica.com/psychiatry.html)

4. Shortly after this fire, another fire broke out in the storage bins where I had stored what was left of my belongings. I have reason to believe that this was NOT a coincidence. (oddly, an operator recently told me that the address of this property does not exist!)

The Alstead, NH Flood and the Rape in 2005;
Near the end of 2005 a flash flood wiped out my Alstead, NH neighborhood. This was reported to have been caused by a stalled storm and a plugged culvert. I now believe that the storm was probably stalled by weather modification technologies and that the culvert may have been intentionally plugged.

After the flood, a couple zoomed in to "help" me, because I no longer had a home to go to. When I got to their home my vehicle was boxed in so that I could not leave without them moving their own vehicles. (There was plenty of room for this to not have to be this way) While I was there I was drugged and raped and lost over a week of time. During my week or two that I was held there I was also brainwashed into thinking that I was responsible for the deaths of the four of my neighbors who died in the flood.

When I later reported the rape to their local police chief, nothing was done about it and the tire on my car suddenly went flat while I was in the police station talking to the chief. He seemed involved and other weird things happened, which I may share later.)

Vandalizing of my RV in Mohave Valley, Arizona in 2011 and 2014;
By this time I had been targeted so much that a beat up RV was all I had left for a home. In 2009 it had been sabotaged, even the fixing of it sabotaged, and I ended up putting it in a storage yard. In the Fall of 2011 I had just gotten a job, and had expressed to someone that I was going to fix my RV as soon as I got enough money saved. Within a couple weeks, I got a call from a police officer, who informed me that my RV had just been vandalized. It was vandalized again in 2014.

Sabotaging of my Vehicles

I have experienced what appears to be remote attacks to my vehicles, which have drained or destroyed batteries and stopped the engine from working, either breaking something or just leaving it un-startable at strategic times. (This was happening severely in 2008, 2009, 2011, 2012, 2013, 2014 and for a period of time in 2015.) My driver side windshield wipers have been repeatedly damaged at eye level. Something has often been sprayed on my windshield that makes it almost impossible to see out of when it gets wet. Some sort of chemical that causes erosion appears to be put on at least two of my vehicles. . .once on my break line which broke while I was driving in 2012. My radios have often been broken or interfered with. Something like salt has been put in my driver side doors through the window opening. I have experienced things being shot into my tires, a screw, metal object and a bullet. I have experienced the sudden unexplainable disintegration of part of the fuel injection system, causing gas to literally pour out into the air filter and onto the ground. I have experienced episodes of unexplainable electrical problems in vehicles. I have experienced repeated episodes of my oil filter being unscrewed. . .and bolts removed from the transmission once. I have experienced an unusual amount of belt breakage and brake losses in vehicles. I have experienced things being moved around inside my vehicle. . .a knife was once placed in it...etc. I now get ongoing threats to disable my vehicle, which is now my home and contains all I have left of my writings. (This is happening again as I write this.)

Laser, Microwave and Psychotronic Weapon Attacks

I have experienced ongoing weapon attacks of various types. Most of it appears to be to interfere with my brain function. Some of it is to inflict physical pain and threaten me with heart attack symptoms and attacks to my lungs and throat. I have had burn marks on my body and have experienced ongoing attacks to my pubic area, severely since the fall of 2013. I have experienced literal brainwashings and attacks to my brain that appear to instigate abnormal feelings of anxiety or anger at strategic times. I almost always hear the ring of microwave weapons in my ears and, aside from the obviously painful attacks, I am periodically inflicted with microwaving that causes things like extreme fatigue, intense heat or sudden bloating of my whole body. The technological targeting sometimes suddenly backs off at times and I have been deeply concerned that this may be done to prevent detection - that those who target me seem to know if possible help is zooming in. It also appears that the opposite happens - it vamps up on interference with my brain, or my writings, at strategic times, in order to make me look crazy or bad at strategic times.

How do I know that these are weapon attacks? Because of the timing of them and the fact that they often come with verbal threats...etc. But, again; with the field of psychiatry being used to inflicting false "mental illness" labels...etc., there is a grave danger for Targeted Individuals – a danger that could have worse results than a physical death. This makes it scary to even report these crimes to agencies that are trained to direct us toward a psychiatrist, in order to prove our sanity - prove that the targeting is really happening. Would you be willing to shift toward proving these crimes through honest medical tests for radiation, cell structure damage, brain damage...etc.?

Of course I realize that "life" happens - that accidents, disasters and deaths naturally happen...etc. But there is so much that has happened to me, which is surrounded by chains of strange and unusual things, that it just can not ALL be cast aside as coincidental. And this list merely hits the tip of the iceberg. Some of it reeks of documented dark/satanic targeting tactics. It appears that someone, had wanted me to be suffering, harmed or dead since the 1970s, and there is no CLEAR and SURE explanation for this. There is, however a few possibilities. But I need protection from further targeting, in order to be able to more fully process this and clarify the details of many things.

Why am I being targeted? I'm not sure. I think it may have started as early as the 1950s with technological experimentation on my mother in Canada. I have also wondered of it may have started with my uncle's family being targeted, because he worked for the FBI and I know of three FBI families who appear to be being targeted. It is possible that I have been being targeted by different groups for different reasons, because it has vamped up at times that make it look like it is because of my writings about embracing feelings and healing instead of taking psychiatric pharmaceuticals.

More can be found in my books and websites below;

"Targeted in America" book; www.poeticpublications.com/booktia5.pdf
"Ramblings of a Targeted Individual" book; www.poeticpublications.com/bookram13.pdf
My Website; www.targetedinamerica.com

P.S. Please keep an open mind when you read my writings. My computers are infiltrated and I even experience interference on library computers. I have recently found four alterations in them and there may be more; The word "not" was erased from this statement in this report, "that they can NOT be viewed as coincidence." Another alteration was in my plea for government help in one of my books; I had written, "Please expose the covert program, and its recruiting process, so that enslaved mind control victims can be set free." and it was altered to, "Please expose the covert program so that we can be set free," which suggests that I am in the program. And the date on the year that Daniel Nadeau died had been changed from 2015 to 2014. The address to www.heartbud.com had been altered. And there may be other things that I have not noticed.

Again; as I send this report I am concerned about the normal process of psychiatric evaluations of victims of these types of crimes. With the field of psychiatry being used to help perform the part of the crimes that forces people to take mind altering drugs after inflicting false "mental illness" labels, which can also lead to the loss of our rights through being declared "incompetent," there is a grave danger for heavily Targeted Individuals – a danger that could have worse results than a physical death. So, I am hoping that you will realize what is happening and not follow the normal protocol of psychiatric evaluations out of respect for our health and safety. Clearly, the only effective method of proving the technological parts of these crimes can be done through honest, unfiltered radio wave detection and medical tests for radiation, cell structure damage, brain damage...etc.

I have faith that there are good people who have been doing what they can, but the infiltration seems strong, the silence still appears to be helping it to grow and <u>I am concerned that the covert</u> <u>"protection" and "rescues" are actually sly enslavement. (Filters may even be built into new</u> <u>technologies, to prevent detection of low frequencies used for mind control.) Please do</u> <u>everything in your power to help expose and stop these crimes so that victims can be set</u> <u>free and have a chance to recover</u>.

Please help us.
www.targetedinamerica.com

Insane

I pray for a world of peace
Love for those who are in need
No one left alone to bleed
I dream. I dream. I dream.
Must be because I am insane.
I see rich people filled with greed
Stealing from those who are in need -
Controlling this crumbling country.
I see. I see I see.
Must be because I am insane.
I see people fighting for their lives
Darkness turning day to night
People thinking its alright.
I cry. I cry. I cry.
Must be because I am insane.
Occults bleeding hearts and souls
Hiding things that we don't know
Evil aiming for control.
I know. I know. I know.
Must be because I am insane.
There are people trying to silence me
In a world that we think is free.
Things I wish I could not see.
I flea. I flea. I flea.
Must be because I am insane.
I pray for a world of peace
Love for those who are in need
No one left alone to bleed
I dream. I dream. I dream.
Must be because I am insane.

Chapter Six

Remedy to Save Humanity

Its Worth Fighting For -

Humanity's Right to Freely Live and Love and Laugh and Cry.

Its Time to Fight Like Heaven!

Personal Growth - A Precious Gift

I feel that the most damaging long term effect, of technological and pharmaceutical interference with our brain's functions, is the destruction of our natural process of personal and spiritual growth. . .and that the dangers of allowing this to continue cannot be overstated. I am deeply concerned that these technological and pharmaceutical intrusions are literally destroying the Heart and Soul of humanity. I feel that the core reason for our lives is to grow into all that we are meant to be and that preventing this growth is extremely damaging on levels that most people may not realize. . .but need to.

The core of the targeting appears to be orchestrated through some sort of satanic occult. Some warn heavily Targeted Individuals to not mention this due to the foolish assumption that if we say we are being targeted by anything satanic, we could be perceived as "mentally ill." But this surely cannot happen through anyone who is wise enough to be aware of the fact that satanic occults do indeed exist and do indeed target people through many deceitful and covert methods. And this devastating reality needs to be exposed and stopped.

Satanic occult members often wear the mask of being kind caring professional community members. . .and are known to move into, or even sometimes run, arenas that offer services for personal or spiritual growth, like churches, yoga centers and mental health facilities...etc. (I have witnessed literal satanic control in all of these arenas.) Add microwave mind control on top of the more well known occult style brainwashing tactics and we have a lethal situation.

The danger in this is immense, because, through these types of arenas, occult members prey on people when they are most vulnerable and impressionable - when they are searching for help. It is a historically documented satanic occult pattern to prey on people who are in a vulnerable state of need for support and love. And this is happening to people.

Sadly, there are even dangers in exposing this, because people need to trust that there are good healthy places, which they can turn to for help and support; places that will not try to brainwash, convert or control them in any way or form. Avoiding seeking help is not good either. It just needs to be done with awareness and caution.

It is vitally important for humanity to listen to the Heart of its own instincts, above all else, through these difficult times we face in our troubled world.

If you cannot find a genuinely supportive arena there are other things you can do to grow and evolve into all that you were meant to be. Perhaps figuring what that is, for yourself, is one of the most growthfull steps you can take.

I have reach the deepest levels of personal growth through private personal/spiritual retreats from society where I write and soul search and can freely allow the full range of feelings, which are sadly becoming non-acceptable in our troubled world, but are needed for our growth.

We do have ourselves and we have a Higher Power that is ALWAYS here for us no matter what. Reach for that Power - let that Love into your Heart and follow it above all else.

Heart Over Mind for Human Kind

P.S. Sadly there are people (like me) who are being heavily targeted in ways that do not allow any sort of private time, or freedom to heal - embrace our feelings, which are desperately needed for rejuvenation or recovery. Instead, we have been being held in a holocaustal prison while being slowly and cruelly destroyed. This hurts indescribably. Lately it's difficult to hold onto hope for my own inner survival, because the targeting has already taken a serious toll on me and it shows no hope of stopping in time for me. But I dream of having the Freedom I need, in order to recover and return to growing into all that I was born to be. I hope you have that Freedom and that you cherish and preserve it with all your mite. It's the most precious thing in life.

It's worth fighting for
Humanity's right to freely live
And love and laugh and cry.
Its time to fight like Heaven!

Remedy to Save Humanity

As I struggle to survive heavy technological, psychological and physical targeting my hope for it's end has searched for remedies, and has found two within my limited wisdom and knowledge.

The most obvious thing is that we need to stop criminal use of all sorts of radio wave technologies and legalize radio wave blockers for heavily targeted citizens until this is accomplished. Please do all that you can to help aid this process. Please find the Heart and the Courage to stand up, expose these crimes, and secure from criminal use, ALL types of radio wave technologies - microwave weapons, laser weapons, psychotronic weapons...etc., which are being used on people. If this can not soon be done the technologies, which are being used to harm us should be disabled, because they are literally destroying the Heart and Soul of humanity.

There are predictions that, around this time, the world will come to an end or some sort of catastrophe will happen. I feel that what has been prophesied has SOME truth to it, but I also feel that this DOES NOT have to happen. I feel that the purpose for foreseeing tragedies is often to offer opportunities for their prevention. And I feel that intentional man-made disasters are NOT "meant to be" and ARE preventable. We CAN prevent Armageddon. But before preventing it we need to face this. . .

HUMANITIES MOST DANGEROUS HOLOCAUST IS ALREADY HAPPENING!

And the Earth being destroyed is NOT the worst possibility! Though I do not believe that the world will come to an end, I feel that something worse will continue happening if the criminal technological targeting of human beings is not stopped. PLEASE think about this! Most mind control victims seem completely unaware of losing control of their own thoughts, feelings and choices in life, which is destroying our natural process of personal and spiritual growth. And the list goes on. . .

The full scope of this seems too horrible to face. But we MUST face it, in order to remedy it. Please face the devastating reality of remote technological targeting of human beings, and help save humanity from the darkness that grows in the shadows of a secret covert war. Please help restore and/or retain our Freedom.

P.S. I'm not a bible thumper. My beliefs lean more toward the practice of opening/healing our Hearts, in order to enable or deepen a direct connection with God. ("God is Love"). But the bible has a lot of good information and it predicted that satin would be tearing families apart and causing horrific levels of disruption in our world...etc., and I feel that this is already happening through criminal use of microwave weapons - mind control technologies. . .and that we must embrace our HEARTS and do all that we can to stop these crimes from continuing and growing. Again. . .predictions can instigate prevention when we let them.

In the deeper past I'd done a lot of writing about our need to heal the Heart of humanity. And I'd done a lot of my own healing work. I am not in very good shape right now, due to being so heavily targeted, but I hope that some of my older publications will touch your heart. Some can be freely downloaded http://www.poeticpublications.com

Working on healing our hearts is almost impossible for those of us who are experiencing heavy targeting/torturing - we are lucky to be just surviving most of the time. And we are in need of protection and safety that is not yet here for us. Please do all that you can to help it come. . .not just for us, but also for the future safety of all of humanity.

As humanity gets brainwashed in various ways, we are being called to listen to our hearts above what is being projected into our minds. Although our Hearts alone can not remedy the technological parts of this crisis, they can help. So it is important that we embrace the depths of our own Hearts and work at healing unresolved issues, which block them. Opening our Hearts to deeper levels will bring more Love into ourselves, our families, our communities. . .our troubled world, and this is desperately needed. Please embrace your Heart to deeper levels than ever before, and help others to do the same.

Love is a weapon which can deflect
Controlling darkness criminals project

Lets Let LOVE Win

I made the following video for us primary Targeted Individuals, but I think it can also apply to the rest of humanity, because there are many who feel too misunderstood and unloved.

This is not a "theory." Its a fight for our lives. Its not a matter of if you "believe it" or not - its a matter of if you are aware and if you can care to help restore our safety and freedom.

World I See

What kind of world can my weary eyes See
What kind of world need grow to be?
A world where kindness picks up paces
To lift broken people from wounded places.
A world where the void of greed and hate
Is filled with Love by the hands of fate,
A world where all is in a state of repair
And none are left in deep despair.

For the Heart of Humanity

No matter what spiritual path, if any, we may choose to follow. . .the most important thing is how open our Hearts are and what effect we have on our fellow human beings. As negative forces push humanity toward suppressing natural feelings with harmful psychiatric pharmaceuticals (like antidepressants), messages which discourage the natural process of grieving and radio wave brainwashings, which aim to turn us into mechanical zombies. . .our Hearts have been closing instead of opening. This is a tragedy that is desperate need of realization and resolution.

The most dangerous outcome of the microwave targeting of humanity is the blocking and/or prevention of our natural process of personal growth. It is imperative that we not only expose and stop these crimes, but also aim to preserve our HEARTS so that we can grow into all that we were meant to be.

In 2004 I launched a mission that aimed to help heal the Heart of humanity. Part of this mission was the public sharing of my own healing process after experiencing unusual chains of difficulties. This took place in seven issues of the Personal Journals, and was targeted so heavily that it was forced to change names and addresses and never had the chance to grow into all that it was meant to be. But what is left of the core of my writings still yearns to reach the Heart of humanity.

I feel like a bit of a hipocrit as I now share this part of my work, because my heart feels beaten up. I'm not in very good shape. I am literally being slowly destroyed and am in desperate need of protection from the technological, chemical and psychological targeting. But when I wrote what I share here my heart was a bit more whole.

The following articles grew from the Personal Journals and continued to strive to reach humanity through two printings of Sharon's Bud and four printings of the Heart Bud. I hope you let them touch your Heart.

Come. . .walk with me. . .into my heart and yours -
Into the places we usually avoid, as we rush through our
Lives in a world that's crying, LOUDER than ever, for us
To slow down, be still, embrace our own Hearts
With one hand, and hold the other out
To our fellow human beings.

The Silent Epidemic

Though most of us have heard that "it's OK to cry," we don't seem to fully realize how incredibly important it is to allow a healthy grieving process after painful situations. We usually close our Hearts, in order to avoid feeling emotional pain. Yet, this closing of our Hearts, no matter how much or how little, is causing even more pain, because crying is what washes away the pain and allows us to feel deeper levels of love and compassion for ourselves and others.

What I call, the "Silent Epidemic" grows and spreads each time we suppress our sadness. The Silent Epidemic is an emotional illness. I know this may sound a bit strange to some of you. But if you read the rest of this, and listen to the Wisdom in your own Heart, I'm sure you'll feel some of the Truth in what I m saying.

Some say that sadness is "negative" or "depressing". Some go so far as to say that it's "un-spiritual" or "dark" to feel, release or express sadness! Some even think that "all we need to do is use our minds to choose joy instead," no matter how we are REALLY feeling! But my experiences show me that this avoidance of our Hearts - this suppression of our sadness, is THE very thing that actually CREATES the "negative" stuff in our world.

I feel certain that humanity's health and well-being depends on each of us allowing the natural cleansing process of healthy grieving, because releasing our emotional pain is what opens our Hearts to deeper levels of Love, Joy and Peace.

We habitually suppress our sadness, because feeling it can be uncomfortable and sometimes overwhelming, especially when it's not supported by the people around us. Even in the most supportive environments, it's difficult to completely embrace grief. Suppression is the easiest route to take, but certainly NOT the healthy one.

Most of us were taught, from the day we were born, to stuff down our feelings of sadness; to "get over it", to pretend it's not there, and "put it behind us" as quickly as possible. Consequently, most of us are better at suppressing than we are at releasing our pain.

We tend to even feel ashamed to go out in public after we've let ourselves deeply cry, because we don't want people to know we've been crying. We act as if crying is doing something wrong or shameful! We waste a lot of energy trying to avoid feeling anything but shallow imitations of joy. We stuff down our sadness with overdoses of caffeine, nicotine, alcohol, food, drugs, pharmaceuticals like anti depressants, TV, sleeping, thinking, working...etc. We tend to keep ourselves so busy and so distracted that there's no time to feel anything! And we often try to stop others from feeling their feelings, because their sadness triggers ours. And on and on and on the unhealthy cycle goes. I feel 100% certain that deeper levels of grieving/crying is an absolute necessity for the health of our Hearts, our families, our communities, our countries. . .our world.

The "Silent Epidemic", is the widest spread, most dangerous epidemic in humanity. No joke! You may think I'm catastrophizing here. But I feel certain that

I'm not. I feel that humanity is at a serious crisis point with this issue. There are far too many things that are pulling us out of our Hearts and preventing our process of personal growth. PLEASE think about this.

Sadness is not depressing!
It's the suppression of it that depresses us.

Suppressing sadness - the closing of our Hearts, appears to be the root cause of ALL the problems humanity faces on both personal and global levels. When we've suppressed too much, it blocks our Hearts - depresses us, or becomes anger that yearns to strike out.

<u>On the smaller scales</u>: not allowing a natural grieving causes our Hearts to start blocking to the point where we also start losing our ability to feel deeper levels of compassion, peace, Love and joy. Greed begins attempting to fill the voids with money and possessions. Our connection to the deeper, wiser parts of ourselves and to the Highest Power, becomes more and more blocked. Is any of this sounding familiar?

<u>On the larger scales</u>: severe suppression of sadness, causes Hearts to become so blocked that they begin filling up with unhealthy levels of greed, warped senses of spirituality, uncontrolled anger or hatred and a thirst for power over others. . .all of which are THE root cause of the destructive wars we experience between family members, religions, cultures, and countries. When Hearts completely block evil moves in.

Now, I'm not suggesting that we walk around trying to cry all the time. But I AM saying that we should work at allowing the depths of our Heart's natural cleansing process - that we should allow and support a healthy grieving process far more than we now do. And I'm praying for us to take a deeper look at the damaging effects of the "NO crying/grieving allowed" messages, we deliver to our children and loved ones. I cringe every time I hear the popular Christmas song, *"You better be good. You better not cry. I'm telling you why. . .Santa Clause is coming to town..."*!!! I'm sure we would not even think of delivering messages like this to our children, if we knew how damaging it is. Sometimes, when I hear this song, I sing along and loudly change the words to, "You'd better cry...", because **our individual Hearts need to utilize their natural cleansing process for our soul's preservation and growth.** The "Silent Epidemic" needs to be cured, in order for us to start healing our world, ESPECIALLY through the tough times we now face. It's OK to cry. It is! It really is.

Crying is like giving the Heart a shower
To wash away accumulated dirt.

~ *Foundation of Humanity* ~

When I step back and look at our troubled world, it appears that the only way we're going to fully heal it is to bring more Love and stability into its foundation, into its roots - into our families.

Within our families we need more support for healthy grieving during times of loss. Within our families we need more of the kind of Love that would not put us down, aim to hurt us or hold us back. Within our families we need less hidden inappropriate behaviors and less mental and emotional abuse. Within our families we need more love, more comfort, more integrity, more compassion - more heart and more support in the process of growing into the wonderfully unique individuals that we all are.

We need our families to be our places of refuge - our safe sanctuaries. Families can't be perfect. But we can do a lot better than what we are doing - we can listen to our Hearts and be here for each other on deeper levels.

In order to have a more positive impact on our world we need MOST of our families to be safe, kind and supportive MOST OF THE TIME. . .and for them to hold a Heart out to those that aren't.

I feel certain that there are far more secretly troubled families than we realize, in EVERY class of society. With what appears to be a satanic occult targeting families with subliminal messaging and brainwshings through our TVs, computers, cell phones and remote microwave technologies, we are in need of drastic counter-measures. Many of us do not seem to realize how wounded our families are becoming.

Within many families there exists a silent rule which says, "it's not OK to face or talk about the damaging things that happen within the family," which prevents healing. Protecting the reputation or appearance of the family is often more important than healing from its mistakes and making things better. Even in the mildest situations, this is damaging.

Please believe that none of this is about judging our families. It's about striving to make things better - it's about healing – its about bringing more Love into humanity through the foundation it is built on.

When we face and heal the painful experiences we stuff into our childhood it lightens our load and frees the future.

Through my efforts to try to understand why some of my own family members treated me as badly as they did, I grew to realize that, within each of their Hearts was a wounded child who needed an outlet for the pain they'd not yet healed. (This was before I realized that a huge part of it is because we are one of many families that has been used for remote technological experimentation.)

I was the scapegoat in a family where denial grew into such damaging levels of mental abuse that I've had to remain almost completely separate from them through most of my adult life. But I still love them.

Although I wish we could all be connected in a good, healthy, healing way I now accept that this will probably never happen.

There are many families who are in this sort of crippled state. And its just too sad that some of us need to protect ourselves from our own families.

But I feel that family members who hurt us deserve as much compassion as we do. We all make mistakes that have ill effects on our children, siblings, parents…etc. And its important for us to face our own mistakes as well as the things we feel hurt by, even if it breaks some dysfunctional family rules, because this is what will help our families heal into a better place.

Pulling family skeletons out of closets may create a bit of chaos, for a while, but when the focus is on healing instead of blaming. . . it is sure to have good results.

Since the mid 1980s, as I aimed to heal from childhood difficulties and then began sharing my healing process in my writings, I've been dealt painful levels of judgment, from all directions. This has been extremely difficult. But in my heart I know that EVERY family has its own share of problems, and that those who leap to point fingers and pass judgment and prevent healing are the ones who have the biggest problems. **NO family is perfect and a lot more healing will take place when we realize this and stop judging each other into hiding our problems and preventing healing.**

Our children Truly are "humanity's future." How can they heal our future until we are healthy enough to keep them whole?

We can't heal the future
Until we feel the past.
We must look behind us
And face the pain at last.
If we want tomorrow
To fill with Love and trust,
We must face the yesterday
That's lost inside of us.
If we wish to open
Our Hearts and truly care,
We must first embrace
The sadness hidden there.

Child I Used To Be

On a lonely summer day I sat at the forest's edge
Feeling the impact of life's hard lessons,
When she came to me, a mere child of three,
In soiled, worn-out clothes and hair of honey gold.
I stared at her in wonder - taking in all I could see,
Realizing that she was the child I used to be.
I thought my eyes deceived until she began to speak -
Glaring at me with big brown eyes, as tears covered her cheeks,
"You spend your life searching but don't remember and see,
That I have been here waiting for you to return to me.
You ran away and forgot the great plans we had for you -
The joyful games we'd play and magical things we'd do."
She sat on the ground rubbing her cold, bare feet
Crying, "You didn't take me with you
To the people we were to meet!
You forgot the castles we were to build in the sand,
And not once did you even TRY to hold my little hand!"
She bowed her head, declaring with a sigh,
"And worst of all, you forgot how to laugh and cry!"
My heart filled with sadness. I knew she was right.
I'd left her to grope alone on a cold and dreary night.
In over twenty years did not return or ever even try
To find the child I cast away. . .for the pain I hid inside.
I reached for her shaking hand and asked if she'd forgive,
While making a sincere promise that, together, we would live.
She climbed into my lap, where we held each other and cried,
Until joy was what was left of the pain we felt inside.

Gifted

A call for us to hold onto, or return to, who we are and what we were born to do with our lives

I feel that every single one of us is gifted in a unique way. Some of us are gifted in earthly ways. Some are gifted in intellectual ways. Some are gifted in creative ways. Some are gifted in spiritual ways...etc. But many of us don't listen to our callings, because its hard to strike out on "the road less traveled," especially when it doesn't look like it will bring the money we view as a symbol of success.

We often fail to realize how important it is for us to honor our own individual Gifts. Consequently, there is very little support for such uniqueness in our copy-cat world. Yet, when we cast aside our own natural Gifts, in order to do what will bring more money or recognition, we also cast aside our own personal power. Its hurts us and stunts our growth.

When we aim for money instead of following our Hearts, we become like sheep running off a cliff, after losing ourselves in greed's engulfing mist.

Have you ever noticed that the one thing, which most historically famous people have in common, is that they passionately put their Hearts and Souls into their work, WITH-OUT following or copying any other human being? History's best writers, scientists, inventors, spiritual leaders, philosophers...etc., reached into the depths of their own Hearts and Souls and used the Gifts they were born with, instead of copying other people's. This is the lesson we need learn.

We are all wonderful. We are all Gifted. And we must reach into our Hearts and Souls, in order to find the Gift that naturally knows.

Those of us who are gifted on spiritual levels can have more difficulty with the process of openly honoring our Gifts because, throughout history, those who are born with deep levels of insightfulness, intuitiveness, healing abilities, prophetic abilities...etc., have been too grossly misunderstood, wrongly labeled and harshly judged. This is sad, because its still happening and hurting innocent people.

At this point in time, there also exists the flip side of this. Since its starting to become "cool" to be a healer or psychic...etc., there are so many wonna-bees jumping into the "spiritually gifted" rolls, that its damaging the credibility of genuine ones, which is destructive for all of us.

The very best we can possibly do with our lives is embrace the Gift we were born with - embrace what came natural to us, when we were children, and use it in our life's work. . .as well as accepting the Gifts in others, even when we don't understand them.

It does not matter if we were born to be a mechanic, a writer, a waitress, a doctor, a politician, a hairdresser, a poet, a farmer, a parent, a psychic, a secretary, a carpenter, a billionaire, an actor, a minister, a prophet...etc. **No purpose is higher than or lower than any other - we all have equal importance. And we are all here to help each other**.

When we put our Hearts into our own Gifts, we put more Love into people's cars, people's food, people's books, people's music, people's lives...etc. Lets do it.

When we use our own Gifts and accept the Gifts in others, we all become more whole and the world becomes more balanced.

What is YOUR natural Gift?

Find it. Embrace it. Use it. Don't lose it.

My Gift

My gift is You. My gift is Me.
It is illuminated in the stars
And travels in your eyes.
My gift rises from
Beneath the largest,
Lonely stone.
It's wings dance
In rays of Light.
My gift shines through
The darkest night.
Its the deepest voice.
In the saddest songs,
We've sung without noise.
My gift is beyond earthly,
Far beyond mundane.
It's wild, free and
Completely untamed.
Like the sun
And full moon,
It's universal,
Yet plays it's own tune.
My gift is all there is,
All that must come to be,
And resides in the depths
Of the Heart in Me.
My gift is Love.

On Suffering

Some of us think that those who have more money are more important or more valuable. Some of us think that we "choose our own reality" or go through tough times because we want to or because we deserve it - some of us judge, assume and blame instead of helping.

Are we letting judgment build arrogance and greed
Into excuses to not help those in need?

It appears so. I feel ashamed to admit that I had moments of believing these sorts of justifications, before I had the rug pulled out from under me and learned (the hard way) that there is a higher purpose to all forms of suffering - one that we often can not fully know or understand.

Sometimes we suffer, in order to experience what we had judged. Sometimes we suffer, in order to gain a deeper appreciation for things we had previously taken for granted. **Sometimes we are cast into struggles, in order to present opportunities for our fellow human beings to open their Hearts and give help that is needed**. Sometimes we are cast into struggles, in order to gain the experience we need, in order to help prevent others from suffering - in order to realize and/or expose a crime or wrong doing. Sometimes...etc.

This list could go on almost endlessly. The reasons for hardship are as vast as the multitudes of complexities in life itself and its not up to us to decide why another person is suffering, especially when that judgment becomes and excuse to not help.

One thing, I'm sure, remains the same:
We are not here for judgmental games.
To grow into Love, is why we came...
And to help find greed's lost shame.

The growing delusion, which claims that "those who are right with God have physical abundance..." IS SIMPLY NOT TRUE. In fact, the opposite is true. Jesus was physically poor, but rich in Heart. There has been a growing degradation of those who "beg for money," although they are merely struggling human beings who are asking for help. Please let your Heart change this.

Reality

I used to think that we, "Create our own reality"
Until Light shone into all I did not see.
There is a grander plan, beyond the sphere of mind,
That sets Wisdom into the toughest hills we climb.
God creates mine.

NATURAL adversity builds strength, but its through LOVE that we Heal.
Adversity, that is intentionally inflicted by fellow human beings, is a crime that needs to be stopped.

Aging Contentment

I stood in the lines where everybody goes
To fix the aging form of skin, hair or nose.
Strong as stone I stood as I studied my reflection
And found these words in my body's deep rejection:
"Each crevice built for tears - these wrinkles on my face,
Are proof of precious years that NOTHING can erase.
In the grey of my fine hair, I sometimes see a glow.
Please handle it with care and let this magic show.
The sparkle in my eyes grows brighter every day.
Please don't cover it up. Don't take THAT all away!
Every blemish, bump or sag, in the eyes of the weak,
May make me a hag. But HEAR these words I speak.
I want to remain human. . .the Truest kind of all.
Don't stretch, tweak or fix me. I don't want to be a doll.
I may not fit in, because of how I feel.
But I don't want to change. So, let me just be REAL."

Can you imagine the extra joy, peace and contentment that will settle into our Hearts when we let go of our foolish concerns about aging and the ridiculous things we do to hide it?

Personalized Disaster Relief

Since the volunteer work I did with Katrina and Rita victims, as well as my own personal experiences with disasters, I've felt saddened by the lack of uncomplicated help for victims. . .and have come to the following conclusion: **When we have the Heart to help, and money to donate, we can help a lot more people with a lot less money when we give directly to individuals and families who are in need.** Through giving DIRECTLY to those in need we can FULLY help dozens of people with the same amount of money that would reluctantly trickle down to partially help just one person. . .through most existing avenues.

Far more disaster victims will be helped when we let our Hearts guide us into the kind of compassion that cares to relieve and prevent further distress by completely lifting people back onto their feet, monetarily, emotionally and mentally. . .instead of offering only enough food or water to sustain their plight. It can cost a lot less to COMPLETELY help people back onto their own feet than it does to keep them down and feed them for long periods of time. . .and its far more humane. **Lets let our Hearts do the deed of FULLY helping those in need.**

After disaster strikes, most victims need these three things, in order to fully recover.

1. A safe, kind, genuinely welcoming place to live. 2. Enough peaceful time to heal, without being pressured to hurry up and "get over it...etc." - time for grieving in an understanding and compassionate environment. **3. Uncomplicated financial assistance,** to lift them back onto their own feet.

Please remember that reluctant help does not feel very helpful, because most disaster victims already feel humiliated and guilty for having to depend on other people, not to mention the state of shock and/or grief they can be experiencing. Victims feel more comfortable with the type of help that comes from GENUINE care. Please find the Heart to care.

Also please remember that interrogating victims, and making them prove their losses, as most agencies do, merely ads to their distress at a time when they are often already so overwhelmed that they have a hard time remembering details, anyway. Its better to just focus on listening to our own instincts - our own Hearts. It is best to just offer a compassionate ear and/or shoulder, instead of suspicion and intense questioning, because they all DO deserve the help they need and should NOT have to be dealt the added distress of having to prove themselves or defend themselves against suspicions, which are usually born from a greedy, selfish search for a reason to not have to help.

NO MATTER WHAT THE SITUATION IS. . .those who continue to suffer just haven't gotten the type of help they need.

"What can I do for you in your time of need?"
(These are the words that plant a healing seed.)

P.S. An old Native American tradition requires that we never let anyone know what we've done to help another person. This is to keep our egos out of it and pull our Hearts into it. Perhaps we can learn something from this wise tradition.

Start Your Own Support Group

Support for the process of healing - feeling/releasing our suppressed grief is desperately needed in our troubled world. And anyone can start a group. Here are some basic guidelines, if you are interested in pulling together a group of friends, neighbors or strangers.

Support Group Guidelines

1. Each member must join with integrity
(Just ONE disrespectful act or comment can make the whole group feel uncomfortable)

2. Make a firm commitment to at least 11 weeks of meetings
(This will allow time to iron out the wrinkles, and then embracing feelings...)

3. Keep the door open to new members
(This keeps it evolving and growing)

4. Treat each other with utmost respect
(It is crucial to NOT sexually approach anyone who is in shock or grief.)

5. Remain consistent with times and dates
(This is important because after a sudden loss or disaster, we need stability and security.)

6. Pass leadership around, so that no one has complete control of the group
(This helps prevent the group from creeping into dysfunctional patterns.)

7. Begin each meeting with some sort of prayer
(or a wish for Healing in your Hearts)

8. Take turns sharing - with only one person talking at a time
(Perhaps use a "Talking Stick")

9. Practice strict levels of confidentiality
(Do not repeat what other members share unless they give their permission.)

10. Encourage talking about losses and painful experiences
(This is what support groups are for.)

11. Focus on fully listening to each individual who speaks
(So that each individual feels heard and cared for)

12. Absolutely NO advising, unless it is specifically asked for
(This is important)

13. Practice the deepest possible levels of compassion
(Try to empathize with each person who shares)

14. Embrace and encourage ALL feelings
(Anger, fear, sadness, joy..., as long as anger is not expressed in hurtful ways)

15. Add any other guidelines or rituals that your group agrees upon
(be open to changes and the needs of every member)

16. End each meeting with a group hug.

Note: Please keep in mind that new losses trigger old ones, so its natural to be suddenly facing childhood trauma or past losses along with the present situation. Embrace it ALL so that healing can happen.

We'll all be happier when its OK to cry.

Lonely Place

Deep inside most Hearts exists a lonely place,
Where sadness hides and silent yearnings
For Love long to be embraced.
This is the place we need to reach -
The depths, where Hearts have much to teach.
But, do we dare reach inside
For sadness that's learned to hide?
Do we dare fully embrace
The tears that long to wash our face?
Do we dare let go of pain
So Love can find its place again?
Perhaps we must.

Heart of Love

The more open our Hearts are - the closer we are to Love/Light/God.
Its an automatic connection. Love is ALWAYS here for us.
All we need do is open our Hearts enough to feel it,
And then keep on opening them until we live it.

Lets let LOVE win

Sight of the Heart Bud
www.heartbud.com

The Heart Bud blog
www.heartbud.blogspot.com

Sharon's Bud
www.sharonsbud.com

Healing the World Begins and Finishes With the Healing of Our Individual Hearts

The End of the Infiltration

Revised March 2016 - global version

A Fictitious Vision of Hope

Once upon a time there were great nations, which stood for Freedom, Liberty and opportunities for people to follow their hearts into making their dreams come true. They were also a safe refuges for those who fled from communism and other sorts of aims for control over citizens. But their open doors left them vulnerable to covert infiltrations, which slyly aimed to destroy them.

By the 1970s, the infiltration was greatly aided by technologies which could remotely shoot brainwashing radio waves into the minds of individuals as well as whole communities as well as government agencies, military barracks and other organizations. Even the weather was being used as a weapon with the help of weather modification technologies.
Radio wave detectors were slyly being made with filters designed to prevent detection of the low frequencies that were being used for mind control and some those who became aware were offered "protection" that was really a sly enslavement. With no defense against technological briainwashings people were being effected in ways that were altering the natural course of their lives and preventing personal and spiritual growth. People's Hearts were being blocked. They often unwittingly followed the brainwashings while thinking that they were their own thoughts and instincts. Confusion and discord spread rapidly, because the programming of peoples minds included severe levels of arrogance, greed, selfishness and immorality.
Before anyone realized the scope of what was happening the nations had begun to crumble. Small businesses were forced into bankruptcy so that large organizations, which were owned by the infiltration, or by those whom they could control, could monopolize the markets. Gas prices soared to aid this process. The military was being taken over. Disasters were wiping out whole communities. The media was being controlled and threatened into not broadcasting anything about it. Families were being torn apart by sadistic covert targeting. Citizens, including government employees, were being roped into covert operations that were secretly run by the infiltrators. Citizens were being pitted against the government and the government was being pitted against citizens. Places where citizens should be able to go for help were being taken over. Eugenics based targetings grew rampant. Many people were being made ill or murdered, with microwave and laser weapons, in ways that looked like natural illnesses and deaths. Mysterious illnesses like lupus, fibromialgia and morgellons spread. Some were murdered with heart attacks, copd, cancer, leukemia and other sorts of lethal illnesses that were being inflicted with microwave weapons. The suffering grew. An uncountable number of lives were being covertly destroyed.
Many people grew scared, because they could sense that something horrible was happening, but they didn't understand what because they had no knowledge of the technologies that were being used on them. Most people's Hearts became blocked and their brains numbed by the combination of psychiatric drugs, which were being pushed onto individuals and even placed in public drinking water, as well as the radio waves (microwaves) that were flooding their communities, homes and minds.
Some people, especially those who became aware of what was happening, were being threatened, stalked and shot and tortured with laser weapons that were built into satellites, but nobody believed them. Some were being diagnosed as "mentally ill" and

institutionalized, although nothing was wrong with them. Some were framed for crimes and incarcerated. Many became homeless (their work and homes sabotaged or destroyed). They suffered indescribably, while their loved ones were manipulated into joining the dark forces against them, mostly in the ways of refusing to help them. The targetings grew into a deadly holocaust.

Terror reined and fear gripped those who did not know how to combat it. The whole situation was looking too hopeless and tears fell from those who cared and wanted to do more than they could to stop the holocaust and end the suffering.

Some uncontrolled Government officials began realizing the scope of what was happening and knew that more needed to be done but were too scared to openly stand up against the crimes, after watching lethal targetings of those who had already tried. In order to save humanity Government officials and media were going to have to publically unite into the grandest stand for Freedom that humanity has ever had to perform. And though it seemed impossible, the time came when a Light shone for them to find the Heart and the courage to take that grand step. And all the world stood still during the first news broadcasts. Their voices echoed through the city streets, homes and businesses around the globe,

"We are experiencing the most difficult crisis that humanity has ever had to face and I must let you know that we are being targeted in ways that are extremely unusual. . . The Freedom of our nations is in serious danger, as is the freedom of citizens and government officials around the globe. Many have already lost their freedom. I want to make you aware of the pharmaceutical and technological targeting so that you have the opportunity to resist it until these crimes can be completely stopped.

It is important that ALL citizens who are members of covert operations immediately stop contact with your leaders, because that organization is the enemy of Freedom. This crisis poses a great need for us to pull together, remain calm and listen to our Hearts above all else. Together, we can restore our Freedom.

We have a few suggestions and hope that you will have some as well; We suggest that neighborhoods form support groups for people to freely express and deal with their concerns and feelings. I hope that targeted families pull together and be witnesses, as well as sources of help, for each other. . .especially for those who have been heavily targeted. Please mail in any suggestions or ideas that you may have on how we can better handle this situation.

Our agencies are probably not going to be able to handle all the phone calls that you may place. But updates will be broadcasted every evening at 7pm and websites have been set up to inform and advise those of you who are experiencing heavy levels of targeting. You will get the protection you need. We are so sorry that this is happening to you. In the event of interference with the broadcasts and websites, news papers will be distributed to trustworthy town and/or state officials who can call meetings for local citizens.

This is a time, like no other, that we are being called to let our Hearts be here for each other in ways that we never have before, because this is not going to be easy. But we can do it - TOGETHER we can do it. We must be sources of help and comfort to each other. Reach out to your fellow human beings with compassion. And stand with me in the trust that WE WILL WIN this battle - WE WILL REGAIN OUR FREEDOM and then we'll help other nations do the same. And we'll do it through letting Love work through our Hearts and spread into our relationships with each other."

When the broadcast ended, people cried and hugged each other. Relief spread across the world, because most people had sensed that something was horribly wrong, but not understanding it had filled them with confusion and fear. Their confusion and fear were now being replaced with understanding and hope. People were resisting the technological mind control as much as possible. Hearts were opening to help those who were in need. Everywhere you looked you there were people helping people.

Within a few months, radio wave blockers and unfiltered radio wave detection devices were legalized and passed out or sold at reasonable prices. Strict regulations were placed on all technologies which emit radio waves and those that could not be regulated were shut down or destroyed. Citizens were no longer being used in covert operations. The lethal targetings stopped. The suffering was over. The nations and their citizens were finally free and on the road to recovery. Peace and Love quickly spread until the whole world became a peaceful heaven on Earth.

World I See

What kind of world can my weary eyes see -
What kind of world need grow to be?
A world where Love fills the void of hate
And freedom is delivered by hands of fate.
A world where everyone picks up paces
To lift broken people from wounded places.
A world where all is in a state of repair
And none are left in deep despair.

Chapter Seven

Cry for Help

Its Worth Fighting For -

Humanity's Right to Freely Live and Love and Laugh and Cry.

Its Time to Fight Like Heaven!

The Bigger Picture

In the beginning of my microwaved grope to figure out the technological targeting, I made a lot of mistakes - the same mistakes that many have been making - mistakes that include pointing fingers at this person or organization or that country or "the government"...etc. Its all just too confusing to fully figure out while still being targeted. But I trust what my heart has felt from the start, especially during those times when I am not being hit too hard with microwaves. . .and am able to step out of my personal situation to look at the bigger picture. This is what I see in that bigger picture;

I see a very dark group of people sitting at the top of a pyramid - people who have been aiming to control the world, including the weather, countries and human beings. They appear to have unlimited resources, are extremely advanced with microwave weapons, mind control technologies - psy-chotronic weapons...etc., and seem to enjoy inflicting pain and suffering and pitting citizens, organi-zations and governments against each other - instigating wars...etc. Beneath them covert wars have been secretly raging through the past few decades. Beneath them whole countries have been being slowly taken over. Beneath them rogue PARTS of citizens, governments, private organizations and medical fields (especially psychiatry) have joined them in eugenics based targetings as well as phar-maceutical and technological mind control on other citizens around the globe. Beneath them healthy people have been being inflicted with various types of physical illnesses, which appear natural but are actually caused by microwave weapons. Beneath them psychiatric pharmaceuticals, which aid technological mind control, have been being heavily pushed and even placed in some public water supplies. Beneath them healthy people have been being falsely labeled as "mentally ill," in order to medicate and/or control them. Beneath them, there has been immense suffering and confusion, especially in those who are being heavily targeted. Beneath them aware victims, who dare to publicly stand up, have been being threatened, fed misinformation and inflicted with technological brain-washings that follow their pattern of manipulating things so that blame is placed in the wrong places. Beneath them humanity has been being enslaved, some have even been heavily targeted/tortured and then "rescued" by their abusers, leaving loved ones to think they are dead. Beneath them are various levels of enslaved mind control victims who are used to control or target others in a covert fight for freedom that is actually destroying freedom. Beneath them there appears to be sly modes of "protection" that actually enslave people and may even offer filtered technologies that prevent detec-tion of frequencies used for mind control. Beneath them unaware, decent people have been used in the foreground of their operations, in order to inconspicuously achieve their aim for control. Beneath them darkness has been spreading like a cancer around the globe. Beneath them humanity has been losing the vital freedom that it needs, in order to think and feel and grow into all that it was meant to be. Beneath them the Heart of humanity has been being slowly destroyed.

This is truly a Holocaustal situation and, within the heartlessness, that it has already inflicted, there is a serious danger that technological mind control could secretly continue (even after the worst of the targeting has stopped) if it is not fully exposed and citizens are not provided with honest, unfiltered methods of radio wave detection and protection until criminal use of the technologies is completely stopped.

But, there is hope, because above those who target and enslave humanity, and those whom they control and use, is a Light that shines for us - a Light that is spreading around the globe and aiming to reach the Hearts of those who can do more to stop the destruction. . .a Light that is begging the Heart of humanity to PEACEFULLY stand up and break the lethal silence, which enables the growth of this holocaust - a Light that longs to protect and comfort tortured victims. . .a Light that calls for the public awareness of the covert program, which can set enslaved victims free, enabling the under-standing that is needed for resistance. . .a Light that shines for all of humanity to understand what has been happening so that recovery can begin.

I understand that my view is not a popular one. But the fact that it has been so heavily targeted validates its truth for me. Over and over again in my writings I have begged for people to stand up, because I have felt (and still feel) that what is needed from aware citizens, especially media and government officials, is HONEST, compassionate, humanitarian types of public stands. . .and good old fashioned "all American" types of public stands, (WHICH ARE NOT LEAD BY THOSE WHO ARE DOING THE TARGETING) against all levels of the technological and pharmaceutical targeting so that our Freedom can be genuinely restored and recovery can begin.

For World Peace

In order for peace to prevail on Earth it must exist between nations, between cultures, between religions and between Governments and Citizens. In order for peace to exist wars must be stoped. And in order for the wars to stop we must find the courage to face and openly acknowledge what is happening in our world today.

Our world is at war - a covert war has been raging against humanity through the past few decades. This war can be more accurately described as a Technological Holocaust. Within our own communities, there are people who are being covertly targeted, people who are victims of technological mind control, people who are being inconspicuously tortured or killed with microwave and laser weapons - people who are suffering in ways that no human being should ever have to endure. And most of these vitims suffer alone. . .while surrounded by people who either do not know or can not care.

In the shadows of this covert war a secret society has been targeting and recruiting citizens around the globe. Some members of this society are abducted through staged deaths; some are tortured and "rescued" into it; some are missing persons or abducted children; some are lured in with money; some are deceived into thinking that it is a good thing; some are blackmailed into obedience; many members are used to help recruit or target fellow citizens; members appear to be technological mind control victims who will probably not want to believe it, but the Truth must be realized, because, in order for peace to prevail humanity must break free from the darkness that aims for control and regain the Freedom to speak, think, dream, feel and Love each other in the ways that God intended.

**Humanity must be free
In order to be all that it can be.**

Never, in the history of humanity, has there been a more crucial time for us to let our Hearts rise into the grandest peaceful stand for Freedom that humanity has ever needed to perform. Lets do it.

Plea for the Heart of Global Governments to Stand Up

There is a desperate need for people to understand the targeting and pull together to comfort and help each other through this crisis, instead of being crushed by it. <u>Please Stop the Covert War and Stand up for Humanity</u>

This is a critical holocaustal situation and I am deeply concerned that if things continue as they are, the worse is yet to come for all of humanity. The technological mind control and enslavement of humanity...etc., needs to be fully exposed and stopped ASAP. **<u>The Heart of humanity needs to stand up and save itself from further destruction</u>**.

It has already been proven that small stands end up being targeted or distracted. What we need is a HUGE stand, one that includes global media and leaders of nations...etc., to unite with a full exposure of all levels of the enslavement and mind control program so that targeted citizens and other government officials can understand and resist and support each other until the targetings are completely stopped. **<u>I pray for leaders of our nations and the UN to let their Hearts stand up for humanity</u>**.

The aim for, and concerns about, lawsuits should be cast away, because this is a time of war and people, in all walks of life, have been being targeted. Government officials and citizens need to be pulling together and standing up instead of fighting against each other.

There is a desperate need for people to understand the targeting and pull together to comfort and help each other through this crisis, instead of being crushed by it. <u>Please Stop the Covert War and Stand up for Humanity</u>

Its Worth Fighting For -

Humanity's Right to Freely Live and Love and Cry and Laugh

Its Time to Fight Like Heaven!

Cry for Global Government Help

Dear UN and Global Government Officials,

Please let your Hearts stand up and save humanity. Please publicly expose the techno-logical and pharmaceutical mind control targeting so that victims and their loved ones can understand what is happening and pull together to support each other until it can be stopped. Please expose the covert program so that enslaved victims can be set free. Please protect from further harm those of us who are being heavily targeted.

The silence is hurting us.

Please do everything in your power to stop criminal use of all sorts of radio wave tech-nologies, pharamceuticals...etc. These things are damaging our brains and preventing our natural process of personal and spiritual growth...etc. Please help save the lives and minds that are being destroyed in eugenics based targetings. Please help free America and other countries that have been being infiltrated. Please help yourselves and the rest of humanity to regain Freedom.

I am still deeply concerned that the covert "rescues" are actually enslave-ment and that technological modes of "protection" could be a sly enslave-ment performed by those who may also be placing filters in detection tech-nologies, in order to prevent detection of the low frequencies that are used for mind control. Help spread the word on this.

Please Help set Humanity Free

P.S. I am concerned about the normal process of psychiatric evaluations of victims of these types of crimes. With the field of psychiatry being used to help perform the part of the crimes that forces people to take mind altering drugs after inflicting false "mental ill-ness" labels, which can also lead to the loss of our rights through being declared "incom-petent," there is a grave danger for heavily Targeted Individuals – a danger that could have worse results than a physical death. So, I am hoping that you will realize what is happening and not follow the normal protocol of psychiatric evaluations out of respect for our health and safety. <u>Clearly, the only effective method of proving the technological parts of these crimes can be done through honest, unfiltered radio wave detection and medical tests for radiation, cell structure damage, brain damage...etc.</u>

World I See

What kind of world can my weary eyes See
What kind of world need grow to be?
A world where kindness picks up paces
To lift broken people from wounded places.
A world where the void of greed and hate
Is filled with Love by the hands of fate,
A world where all is in a state of repair
And none are left in deep despair.

www.heartbud.com

www.targetedinamerica.com

I am still deeply concerned that the covert "rescues" are actually enslavement and that technological modes of "protection" could be a sly enslavement performed by those who may also be placing filters in detection technologies, in order to prevent detection of the low frequencies that are used for mind control. Help spread the word on this.

There is one copy of this book, which was purchased and may have been altered. My computers are infiltrated and the file in my computer was replaced with another one, as I aimed to upload the sixth edition onto my publishing site. On the day that this happened a book sold. A tear was erased from from the picture on this publication in some previous editions. Its back to having it's seven tears.